Springer Texts in Electrical Engineering

Frank M. Callier
Charles A. Desoer

Multivariable Feedback Systems

Consulting Editor: John B. Thomas

With 20 Illustrations

Springer-Verlag
New York Heidelberg Berlin

A Dowden &
Culver Book

Frank M. Callier
Department of Mathematics
Facultés Universitaires
Notre-Dame de la Paix
Namur,
Belgium

Charles A. Desoer
Department of Electrical Engineering
and Computer Sciences
University of California
Berkeley, California 94720
U.S.A.

Library of Congress Cataloging in Publication Data
Callier, F. M.
Multivariable feedback systems.
(Springer texts in electrical engineering)
Includes index.
1. Feedback control systems. I. Desoer,
Charles A. II. Title. III. Series.
TJ216.C34 1982 629.8′3 82-10407

Printed in the United States of America.

9 8 7 6 5 4 3 2 1

ISBN 0-387-90768-8 Springer Verlag New York Heidelberg Berlin (hard cover)
ISBN 3-540-90768-8 Springer-Verlag Berlin Heidelberg New York (hard cover)
ISBN 0-387-90759-9 Springer-Verlag New York Heidelberg Berlin (soft cover)
ISBN 3-540-90759-9 Springer-Verlag Berlin Heidelberg New York (soft cover)

Preface

This volume is the result of our teaching in the last few years of a first-year graduate course on multivariable feedback systems addressed to control engineers. The prerequisites are modest: an undergraduate course in control (for acquaintance with concepts, terms, and design goals) and a senior-graduate course in linear systems. This volume covers lumped linear time-invariant multi-input multi-output systems with strong emphasis on control problems. The purpose is to provide a rapid introduction to some of the main and simpler results of control theory and to provide access to the current literature. Note that our exposition pays particular attention to the time-domain behavior of the systems under study. Note also that we cover neither optimization nor stochastic systems since these topics are treated in separate courses.

As is obvious from its abundant literature, multivariable control is a very rapidly developing field. Consequently, we have no expectation that our exposition will become definitive; however, we hope that our efforts will be found useful.

To get an idea of the contents, we suggest reading carefully the table of contents and the introduction of the chapters. Roughly, Chapter 1 is an introduction to feedback issues in a multivariable context (desensitization, large gain, singular values, etc.). Chapters 2 and 3 cover the mathematical tools for handling transfer functions as polynomial-matrix fractions and for studying systems described by polynomial matrices. Chapter 4 uses these tools to cover the general theory of interconnected systems. The last four chapters deal with feedback control theory: compensator design for controlling the dynamics, for tracking, and special design techniques for stable plants. Chapter 5 covers some of these topics in the simple single-input single-output context. The first two appendices briefly cover useful algebraic topics. The last appendix covers the division of a polynomial vector by a polynomial matrix.

In the course of developing this material we benefited from many discussions from coworkers and colleagues, and good questions from sharp-eyed students: we are deeply grateful to all of them.

We gratefully acknowledge the support over the years of the Belgian National Fund of Scientific Research, the National Science Foundation, the Joint Services Electronics Program, the Facultés Universitaires de Namur, and the University of California at Berkeley.

Our special thanks go to our wives Nicole and Jackie and our children Ann and Mike for their support during the long hours of writing.

We are very grateful to Doris Simpson for her expert typing.

Berkeley, August 1, 1981.

Contents

Chapter 1. On the Advantages of Feedback 1
1.1. Introduction 1
1.2. Singular Value Decomposition of a Matrix 2
1.3. Large Loop Gain 10

Chapter 2. Matrix Fraction Description of Transfer Functions 17
2.1. Introduction 17
2.2. Polynomials, Euclidean Rings, and Modules 18
2.3. Polynomial Matrices 24
2.3.1. Divisors, Coprimeness, Rank 24
2.3.2. Elementary Operations on Polynomial Matrices 26
2.3.3. Elementary Operations and Differential Equations 29
2.3.4. Standard Forms: Hermite and Smith Forms 32
2.3.5. The Solution Space of $D(p)\,\xi(t) = \theta \quad t \geq 0$ 41
2.3.6. Greatest Common Divisor Extraction 51
2.4. Matrix Fraction Descriptions of Rational Transfer Function Matrices 56
2.4.1. Coprime Fractions 56
2.4.2. Smith-Mcmillan Form; Relation to Coprime Fractions 62
2.4.3. Proper Transfer Function Matrices 67
2.4.4. Poles and Zeros 73
2.4.5. Dynamical Interpretation of Poles and Zeros 82
2.5. Realization and Polynomial Matrix Fractions 87

Chapter 3. Polynomial Matrix System Descriptions and Related Transfer Functions 92
3.1. Introduction 92
3.2. Dynamics of a PMD; Redundancy 92
3.2.1. Dynamics of a PMD 92
3.2.2. Reachability of PDMs 97
3.2.3. Observability of PMDs 101
3.2.4. Minimality, Hidden Modes, Poles, and Zeros 104
3.3. Well-Formed and Exponentially Stable PMDs 107
3.3.1. Well-Formed PMDs 107
3.3.2. Exponentially Stable PMDs 121
3.4. Transfer Functions: Right-Left Fractions; Internally Proper Fractions 131

Chapter 4. Interconnected Systems 140
4.1. Introduction 140
4.2. Exponential Stability of an Interconnection of Subsystems 140
4.3. Feedback System Exponential Stability 153
4.4. Special Properties of Feedback Systems 159

Chapter 5. Single-Input Single-Output Systems 163
5.1. Introduction 163
5.2. Problem Statement and Analysis 163
5.3. Design 175

Chapter 6. The Closed-Loop Eigenvalue Placement Problem 181
6.1. Introduction 181
6.2. The Compensator Problem 181

Chapter 7. Asymptotic Tracking 196
7.1. Introduction 196
7.2. Theory of Asymptotic Tracking 197
7.3. The Tracking Compensator Problem 212

Chapter 8. Design with Stable Plants 219
8.1. Introduction 219
8.2. Q-Parametrization Design Properties 219
8.3. Q-Design Algorithm for Decoupling by Feedback 229
8.4. Two-Step Compensation Theorem for Unstable Plants 232

Epilogue 243

Appendices 245
A. Rings and Fields 245
B. Matrices with Elements in a Commutative Ring $\mathbb{K}$ 249
C. Division of a Polynomial Vector on the Left by a Polynomial Matrix 253

References 260

Symbols 264

Subject Index 270

Note to the Reader

A course that would cover the essentials would consist of the following: Chapter 1; Sec. 2.3 and Secs. 2.4.1 to 2.4.4; Secs. 3.2.1, 3.2.2, 3.2.3 (results only), 3.3.1 (results), 3.3.2, and 3.4 (results); Chapter 4; Chapter 5; for the last three chapters the introductions provide guidance to the topics of interest.

We have not hesitated to lighten our style by using a number of mathematical symbols, e.g., $\mathbb{C}_+$ for the closed right-half of the complex plane; these symbols are almost standard in the literature. For convenience a table of symbols is placed ahead of the subject index (p. 264).

We also use a number of abbreviations, e.g., PMD for polynomial matrix system description; they are included at their lexicographic locations in the subject index.

Note to the Reader

Chapter 1. On the Advantages of Feedback

1.1. Introduction

Feedback is a major engineering invention. The main reasons for using feedback are:

1. Desensitization of the closed-loop system performance due to plant variations (see Sec. 1.3).
2. Reduction of the closed-loop system response due to external disturbances or noise (see Sec. 1.3).
3. Robust asymptotic tracking and robust disturbance rejection for certain classes of inputs (see Sec. 5.2 and 7.2).
4. Improvement of dynamic response: e.g., stabilization of unstable or insufficiently stable plants (see Secs. 5.2 and 6.2 and Chapter 8).
5. Achieving a closed-loop system that is more linear than the plant (e.g., hi-fi amplifiers, measuring equipment, op-amps).
6. Positive feedback is used to create instability (e.g., oscillators, flip-flops, Q-multipliers).

This chapter investigates the conditions under which large loop gain achieves feedback objectives 1 and 2.

For multiple-feedback-loop systems, the notion of large loop gain needs a clear definition. It turns out that the singular value decomposition of a matrix is an appropriate tool for describing this notion.

Thus Sec. 1.2 describes the singular value decomposition of a complex matrix (Theorem 1.2.21), and develops the idea of directional gain (Theorem 1.2.57).

Section 1.3 studies closed-loop exponentially stable feedback systems in the frequency domain under large loop gain. The notion of large loop gain is defined and its effects are investigated (a) for response shaping and desensitization of the closed-loop output under external disturbances (Theorem 1.3.24), and (b) for desensitization under plant variations (Theorems 1.3.35 and 1.3.43).

Although the analysis in Sec. 1.3 is carried out algebraically in the frequency domain, there is everywhere an implied time-domain interpretation. In particular, with $R(0)$, viz., the ring of exponentially stable transfer functions, given by

$$R(0) := \{f \in \mathbb{R}_p(s) : f \text{ is analytic in } \mathbb{C}_+\},$$

closed-loop exponential stability will mean here that all closed-loop transfer functions have entries in $R(0)$, or equivalently are exponentially stable. We shall then assume that all open-loop transfer functions are generated by an underlying time-domain model, called PMD (see Sec. 3.2), which is well-formed and has no unstable hidden modes (see Sec. 3.3). Consequently, the notion of closed-loop exponential stability will coincide with the notion of exponential stability of the underlying closed-loop PMD (see Secs. 3.3 and 4.2): the latter notion has time-domain implications given by Theorem 3.3.2.19.

1.2. Singular Value Decomposition of a Matrix

Singular value decomposition (s.v.d.) is an algorithm which determines the singular values of a matrix $A \in \mathbb{C}^{m\times n}$. The singular values allow one to determine the size of the action of the underlying linear map in all directions. As a consequence it will be possible to express mathematically that the action is small or large in all directions ("small" or "large" gain).

The exercises that follow delineate useful properties of any matrix $A \in \mathbb{C}^{m\times n}$.

1 Exercise [Orthogonal decomposition of domain and codomain]
Show that if $A \in \mathbb{C}^{m\times n}$ with rank r,

2 $\quad \mathbb{C}^n = \text{dom}(A) = R(A^*) \overset{\perp}{\oplus} N(A),$

3 $\quad \mathbb{C}^m = \text{codom}(A) = R(A) \overset{\perp}{\oplus} N(A^*),$

4 $\quad \text{rk } A = \text{rk } A^* = r.$

5 Exercise [Bijections induced by orthogonal decomposition]
Show that if $A \in \mathbb{C}^{m\times n}$ with rank r, then with $A|_{R(A^*)}$ and $A^*|_{R(A)}$ denoting the restrictions of A on $R(A^*)$ and A^* on $R(A)$, resp.:

6 $\quad A|_{R(A^*)}: R(A^*) \to R(A)$ is a bijection,

7 $\quad A^*|_{R(A)}: R(A) \to R(A^*)$ is a bijection,

8 $\quad R(A^*) = R(A^*A)$ and $N(A) = N(A^*A)$,

9 $\quad R(A) = R(AA^*)$ and $N(A^*) = N(AA^*)$,

10    $rk(AA^\star) = rk(A) = rk(A^\star) = rk(A^\star A) = r$.

11   Comment. Exercises 1 and 5 show that any linear map represented by a matrix $A \in \mathbb{C}^{m \times n}$ can be made invertible by restricting its domain and codomain. The same holds for $A^\star \in \mathbb{C}^{n \times m}$. The situation is displayed by Fig. 1.

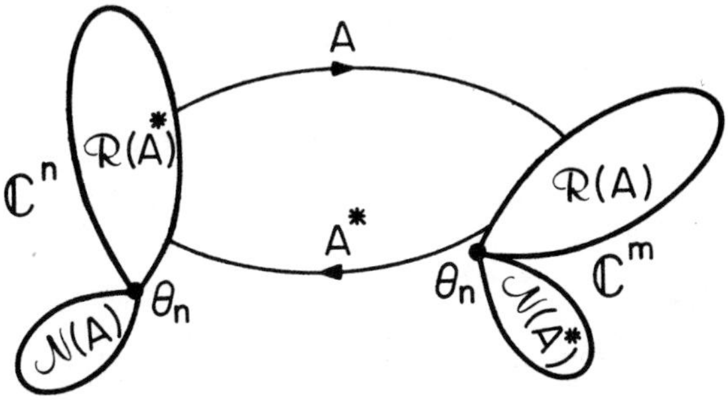

Fig. 1.   Bijections associated with $A \in \mathbb{C}^{m \times n}$ through
orthogonal decomposition of domain and codomain.

S.v.d. will display bijections (6) and (7) in a simple diagonal form by the choice of orthonormal bases for $R(A^\star)$ and $R(A)$. We note further that (a) maps (6) and (7) are adjoints of each other (indeed, $\forall x \in R(A^\star)$ and $\forall y \in R(A)$ $\langle Ax, y \rangle = \langle x, A^\star y \rangle$); and (b) the composition of maps (6) and (7) delivers two bijections $A^\star A|_{R(A^\star)} : R(A^\star) \to R(A^\star)$ and $AA^\star|_{R(A)} : R(A) \to R(A)$ which are Hermitian positive definite, having as eigenvalues the nonzero eigenvalues of resp. $A^\star A$ and $AA^\star$. We have also

15   Exercise [Common nonzero eigenvalues of $A^\star A$ and $AA^\star$]. Let $A \in \mathbb{C}^{m \times n}$. Show that for the Hermitian positive semidefinite matrices $A^\star A \in \mathbb{C}^{n \times n}$ and $AA^\star \in \mathbb{C}^{m \times m}$. $\forall$ nonzero $\sigma \in \mathbb{R}$,

16    $\det \begin{bmatrix} \overset{m}{\sigma I_m} & \overset{n}{A} \\ A^\star & \sigma I_n \end{bmatrix} \begin{matrix} m \\ n \end{matrix} = \sigma^{m-n} \det[\sigma^2 I_n - A^\star A] = \sigma^{n-m} \det[\sigma^2 I_m - AA^\star]$,

17 A^*A and AA^* have identical nonzero eigenvalues $\sigma^2 > 0$.

18 Comment. The s.v.d. of $A \in \mathbb{C}^{m\times n}$ will display the square roots of these eigenvalues: these are called positive singular values of A.

21 Theorem [Singular value decomposition of a complex matrix]. Let $A \in \mathbb{C}^{m\times n}$ be of rank r. Then there exist matrices $U \in \mathbb{C}^{m\times m}$, $V \in \mathbb{C}^{n\times n}$, $\Sigma_1 \in \mathbb{R}^{r\times r}$ s.t.

(a)

22 $$V = \underbrace{\left[\, V_1 \;\vdots\; V_2 \,\right]_n}_{r \quad n-r} \in \mathbb{C}^{n\times n}$$

satisfies

V is unitary, i.e., $V^*V = I_n$,

$\mathcal{R}(V_1) = \mathcal{R}(A^*)$,

The columns of V_1 form an orthonormal basis for $\mathcal{R}(A^*)$,

$\mathcal{R}(V_2) = \mathcal{N}(A)$,

The columns of V_2 form an orthonormal basis for $\mathcal{N}(A)$,

The columns of V form a complete orthonormal basis of eigenvectors of A^*A.

(b)

23 $$U = \underbrace{\left[\, U_1 \;\vdots\; U_2 \,\right]_m}_{r \quad m-r} \in \mathbb{C}^{m\times m}$$

satisfies

U is unitary, i.e., $U^*U = I_m$,

$\mathcal{R}(U_1) = \mathcal{R}(A)$,

The columns of U_1 form an orthonormal basis for $\mathcal{R}(A)$,

$\mathcal{R}(U_2) = \mathcal{N}(A^*)$,

The columns of U_2 form an orthonormal basis for $N(A^*)$,

The columns of U form a complete orthonormal basis of eigenvectors of AA^*.

(c) Under the vector representations for $\mathcal{R}(A^*)$ and $\mathcal{R}(A)$ given by

24 $$\mathbb{C}^r \to \mathcal{R}(A^*) : \xi^1 \mapsto x^1 = V_1\xi^1, \quad \mathbb{C}^r \to \mathcal{R}(A) : \eta^1 \mapsto y^1 = U_1\eta^1,$$

the bijections induced by orthogonal decomposition, viz.,

25 $$A|_{R(A^*)} : R(A^*) \to R(A), \quad A^*|_{R(A)} : R(A) \to R(A^*)$$

have representations

26 $$\Sigma_1 : \xi^1 \mapsto \eta^1 = \Sigma_1\xi^1, \quad \Sigma_1 : \eta^1 \mapsto \xi^1 = \Sigma_1\eta^1,$$

with

27 $$\Sigma_1 := \mathrm{diag}[\sigma_1, \sigma_2, \cdots, \sigma_r] \in \mathbb{R}^{r\times r},$$

s.t. $$\sigma_1 \geq \sigma_2 \geq \cdots \geq \sigma_r > 0,$$

where the σ_i, $i \in \underline{r}$, are the square roots of the common nonzero eigenvalues of A^*A and AA^*, called positive singular values of A.

(d) $A \in \mathbb{C}^{m\times n}$ has a dyadic expansion

30 $$A = U_1\Sigma_1V_1^* \text{ or equiv. } A = \sum_{i=1}^{r} \sigma_i u_i v_i^*$$

where the u_i, v_i for $i \in \underline{r}$ are columns of U_1, resp. V_1.

(e) $A \in \mathbb{C}^{m\times n}$ has a singular value decomposition (s.v.d.)

31 $$A = U\,\Sigma\,V^*,$$

where

32 $$\Sigma := \begin{bmatrix} \Sigma_1 & 0 \\ 0 & 0 \end{bmatrix} \begin{matrix} r \\ m-r \end{matrix} \quad \text{(columns: } r,\ n-r\text{)}.$$ ∎

37 Comments. (a) If $A \in \mathbb{R}^{m\times n}$, then $U \in \mathbb{R}^{m\times m}$ and $V \in \mathbb{R}^{n\times n}$ are orthogonal ($U^TU = I_m$, etc.) and define "rotations" on codomain and domain.

(b) Theorem 21 teaches us that, modulo unitary transformations in the domain and codomain, matrix $A \in \mathbb{C}^{m\times n}$ can be diagonalized displaying the bijections induced by orthogonal decomposition of Exercise 5 in their simplest diagonal form: the positive singular values so displayed are the square roots of the nonzero eigenvalues of both AA^* and A^*A (see Comments 11 and 18).

(c) Appropriate references for the theory are [Ste.1]; for programs and subroutines, see [Gar.1].

(d) S.v.d. has the following geometric content: from (22), (23), and (31), $A[V_1 \vdots V_2] = [U_1\Sigma_1; 0]$, whence for the "active part" $AV_1 = U_1\Sigma_1$, i.e., $Av_i = u_i\sigma_i$ for $i \in \underline{r}$: <u>any linear map "rotates"</u> ($v_i \mapsto u_i$, $i \in \underline{r}$) <u>and "scales"</u> ($u_i \mapsto u_i\sigma_i$, $i \in \underline{r}$) (the v_i's and u_i's are columns of V_1 and U_1).

40 <u>Proof of Theorem 21</u>. By construction.

(a) $A \in \mathbb{C}^{m\times n}$ has rank r, whence by (10) the Hermitian p.s.d. matrix A*A has rank r with n nonnegative eigenvalues σ_i^2, $i \in \underline{n}$, ordered as

41 $$\sigma_1^2 \geq \sigma_2^2 \geq \cdots \geq \sigma_r^2 > 0 = \sigma_{r+1}^2 = \cdots = \sigma_n^2,$$

to which correspond a complete orthonormal eigenvector basis $(v_i)_1^n$ of A*A. This ordered set of $\mathbb{C}^n$ vectors defines a unitary $n \times n$ matrix V having a partition and properties (22) using (8).

(b) The $m \times m$ matrix U is now constructed as follows. Define

42 $$\Sigma_1 := \text{diag}[\sigma_1, \sigma_2, \cdots, \sigma_r],$$

where the σ_i, $i \in \underline{r}$, are those given in (41). From properties (22), especially V_1 is a complete set of orthonormal eigenvector for the nonzero eigenvalues of A*A, $A^*AV_1 = V_1\Sigma_1^2$, whence $(AV_1\Sigma_1^{-1})^*(AV_1\Sigma_1^{-1}) = I_r$. This defines a $m \times r$ matrix

43 $$U_1 := AV_1\Sigma_1^{-1}$$

and from the above we obtain by calculation

$$U_1^*U_1 = I_r$$
$$AA^*U_1 = U_1\Sigma_1^2.$$

Since A*A and AA* have identical nonzero eigenvalues (hence the diagonal elements of Σ_1^2 include <u>all</u> the positive eigenvalues of AA*), it follows that the columns of U_1 form an orthonormal basis for $R(AA^*) = R(A)$: properties (23^2) and (23^3) hold.

Define now any $m \times (m-r)$ matrix U_2 with <u>orthonormal</u> columns s.t. $U_2^*U_1 = 0$. Note that

$$U = \Big[\underbrace{U_1}_{r} \vdots \underbrace{U_2}_{m-r}\Big] m \in \mathbb{C}^{m\times m}$$

is a unitary matrix confirming (23^1); moreover, $\mathbb{C}^m = R(U_1) \overset{\perp}{\oplus} R(U_2)$. Now by (3), $\mathbb{C}^m = R(A) \overset{\perp}{\oplus} N(A^*)$, where we know that $R(U_1) = R(A)$: hence $R(U_2) = N(A^*)$ and (23^4) and (23^5) follow. Notice also that by (9) $N(A^*) = N(AA^*)$. Therefore, property (23^6) follows since U is unitary, $AA^*U_1 = U_1\Sigma_1^2$ and $R(U_2) = N(AA^*)$.

(c) The proof of assertion (c) uses the relations

44 $$AV_1 = U_1\Sigma_1 \qquad A^*U_1 = V_1\Sigma_1$$

which follow from (43); it also uses the fact that the nonzero eigenvalues of A*A in (41)-(42) are also those of AA*.

(d) The dyadic expansion (30) follows from (44).

(e) The singular value decomposition (31)-(32) follows because

$$A[V_1 \mid V_2] = [U_1\Sigma_1 \mid 0] = [U_1 \mid U_2]\Sigma.$$ ∎

For motivating the definition and results below, we have

48 Exercise. Show that a unitary transformation $x = V\xi$ (with $V^*V = I_n$) leaves the Euclidean norm unchanged, or equiv. $\|x\| = \|\xi\|$.

49 Exercise. Let $A \in \mathbb{C}^{m\times n}$ have a s.v.d. (31)-(32) and apply the unitary transformations $x = V\xi$, $y = U\eta$, where V and U are given by (22) and (23). Then

50 $$\|Ax\| = \|\Sigma\xi\|.$$

Hence for $x = v_i$ (hence $\xi = (0, \cdots, 0, 1, 0, \cdots, 0)^T$) with $i \in \underline{n}$

51 $$\|Av_i\| = \sigma_i \qquad \forall i \in \underline{n},$$

where the σ_i are the square roots of the n eigenvalues of A*A.

52 Comment. Note that the n square roots of the eigenvalues σ_i^2 of A*A measure the size of the actions of A on a complete set of orthonormal directions of the domain of A. For this reason we have

53 Definition. Let $A \in \mathbb{C}^{m\times n}$ have rank r. Then the n nonnegative square roots σ_i of the eigenvalues of A*A are called singular values of A. When ordered according to

$$\sigma_1 \geq \sigma_2 \geq \cdots \geq \sigma_2 > 0 = \sigma_{r+1} = \cdots = \sigma_n$$

the first r singular values are called positive and are given by the s.v.d. (31)-(32).

A complete picture of the action of a matrix $A \in \mathbb{C}^{m\times n}$ is the image under A of the unit sphere of $\mathbb{C}^n$. In fact, this combines display of directional influence and size of action.

57 Theorem [Indicator ellipsoid of a matrix $A \in \mathbb{C}^{m\times n}$]. Let $A \in \mathbb{C}^{m\times n}$ be of rank r and have a s.v.d. (31)-(32). Consider the domain and codomain coordinate transformations defined by $x = V\xi$ and $y = U\eta$ with U and V defined by (22) and (23).

Let S_n denote the unit sphere of $\mathbb{C}^n$ centered at θ_n, equiv.

58 $$S_n := \{x \in \mathbb{C}^n : \|x\| = 1\}.$$

Let $A[S_n]$ denote the image under A of S_n. U.t.c. with $\eta = (\eta_i)_1^m \in \mathbb{C}^m$,

59 $$r = m = n \Rightarrow A[S_n] = \{y \in \mathbb{C}^m : y = U\eta,\ \sum_{i=1}^{r} (\eta_i/\sigma_i)^2 = 1\},$$

60 $$r = m < n \Rightarrow A[S_n] = \{y \in \mathbb{C}^m : y = U\eta,\ \sum_{i=1}^{r} (\eta_i/\sigma_i)^2 \leq 1\},$$

61 $$r = n < m \Rightarrow A[S_n] = \{y \in \mathbb{C}^m : y = U\eta,\ \sum_{i=1}^{r} (\eta_i/\sigma_i)^2 = 1,\ \eta_{r+1} = \cdots = \eta_m = 0\},$$

62 $$r < \min(m,n) \Rightarrow A[S_n] = \{y \in \mathbb{C}^m : y = U\eta,\ \sum_{i=1}^{r} (\eta_i/\sigma_i)^2 \leq 1,$$
$$\eta_{r+1} = \cdots = \eta_m = 0\}.$$

63 Comment. A maps the unit sphere of $\mathbb{C}^n$ onto an r-dimensional ellipsoid in $R(A)$ with as principal axes the columns u_i of U of length $\sigma_i > 0$ for $i \in \underline{r}$: the points in $R(A)$ interior to the ellipsoid have to be included iff A has not full column rank: the ellipsoid has m principal axes with positive length iff A has full row rank. Note that the four cases in (59)-(62) are mutually exclusive. Notice also that lengths of the principal axes are the positive singular values and that the relationship between the principal axes for S_n and the ellipsoid is

$$Av_i = \sigma_i u_i \qquad i \in \underline{r},$$

where the u_i and v_i are columns of U_1 and V_1: see the s.v.d. of A.

66 Proof of Theorem 57. We prove only the completely degenerate case $r < \min(m,n)$. Since $\mathbb{C}^n = R(A^*) \overset{\perp}{\oplus} N(A)$, it follows that $A[S_n]$ is the image under A of the orthogonal projection of S_n onto $R(A^*)$ which is B_r, viz., the unit ball of $R(A^*)$ centered at θ_n. Under the coordinate transformation $x = V\xi$, where V is given by (22), we have, with $\xi = (\xi_i)_{i=1}^n$,

$$B_r = \{x \in C^n : x = V\xi, \sum_{i=1}^{r} \xi_i^2 \leq 1, \xi_{r+1} = \cdots = \xi_n = 0\}$$

and

$$A[S_n] = A[B_r] \subset R(A).$$

Now using the s.v.d. $AV = U\Sigma$ and $y = U\eta$ with U given by (23), it follows that $\eta = \Sigma\xi$ with $\xi_i = \eta_i/\sigma_i$ for $i \in \underline{r}$ and $\eta_{r+1} = \cdots = \eta_m = 0$: (62) holds. ∎

67 Exercise. Prove assertions (59)-(61) of Theorem 57.

We are now in a position to discuss norm implications. Note that for any matrix $A \in \mathbb{C}^{m\times n}$ the ℓ_2-induced norm of A is given by

68 $$\|A\| = \max_{x\neq\theta} \{\|Ax\|/\|x\|\} = \max\{\|Ax\| : \|x\| = 1\}.$$

Hence

$$\|A\| = \max\{\|y\| : y \in A[S_n]\}:$$

<u>the norm is the size of longest vector of the image under A of the unit sphere of $\mathbb{C}^n$.</u>

71 Theorem [Norm of a matrix and of its inverse]. Let $A \in \mathbb{C}^{m\times n}$ be of rank r and have a s.v.d. (31)-(32). In particular, A has n singular values

72 $$\sigma_1 \geq \sigma_2 \geq \cdots \geq \sigma_r > 0 = \sigma_{r+1} = \cdots = \sigma_n.$$

U.t.c.

(a)

73 $$\|A\| = \max\{\|Ax\| : \|x\| = 1\} = \sigma_1$$

(b)

74 $$\min\{\|Ax\| : \|x\| = 1\} = \sigma_n,$$

where

75 $$\sigma_n > 0 \Leftrightarrow A \text{ has full column rank.}$$

(c) If A is square and nonsingular ($r = m = n$), then

76 $$\|A^{-1}\| = 1/\sigma_n.$$

77 Proof. Exercise: use Theorem 57 and the norm invariance of unitary transformations.

78 Comments. (a) The largest singular value is the norm of A and the smallest singular value is the inverse of the norm of A^{-1}.
(b) In the square nonsingular case σ_n is a worst-case sensitivity parameter for the equation $Ax = b$, $b \in \mathbb{C}^n$. Perturb b by δb and calculate δx: $\delta x = A^{-1}\delta b$. If $\delta b = \delta\alpha\, u_n$, where $\delta\alpha$ is scalar and u_n is the nth column of U of (23), then $\|\delta x\| = |\delta\alpha|\, 1/\sigma_n$ (thus when $\sigma_n << 1$, a small δb can cause a very large δx). Using the indicator ellipsoid of A^{-1}, it is seen that b-perturbations in any other direction cause smaller x-perturbations.
(c) If we want the size of the action of A to be large (small) in all directions, we must require σ_n to be large (σ_1 to be small resp.).

1.3. Large Loop Gain

In this section we consider a multivariable feedback system Σ given by Fig. 1 under the following assumptions.

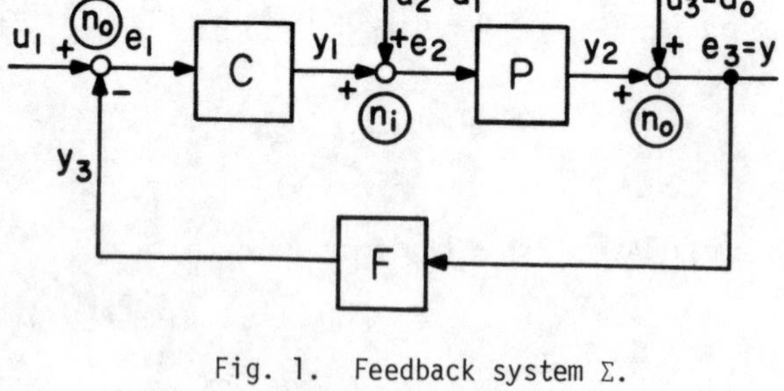

Fig. 1. Feedback system Σ.

1 Assumption 1. P is the plant and C and F are resp. the precompensator and feedback compensators described by their transfer functions

$$P \in \mathbb{R}_{p,o}(s)^{n_o \times n_i}, \; C \in \mathbb{R}_p(s)^{n_i \times n_o}, \; F \in \mathbb{R}_p(s)^{n_o \times n_o},$$

where it is assumed that P, C, and F have underlying well-formed PMDs with no unstable hidden modes (see Chapter 3). ∎

Note that each subsystem is preceded by a summing node with exogenous inputs u_1, $u_2 = d_i$, $u_3 = d_o$, which are the closed-loop system input, and additive disturbances at the input and output of the plant; e_i and y_i for $i = 1 \sim 3$ are the subsystem input and outputs; in particular $e_3 = y$ is the closed-loop system output.

2 Assumption 2. Feedback system Σ is exp. stable, or equiv. the closed-loop transfer functions $H_{y_j u_i} : u_i \to y_j$, for i and j = 1 ∼ 3, are exp. stable (equiv. $\in E(R(0))$; see (2.2.6)). ∎

3 Comment. By the methods of Chapter 4 it can be shown that under Assumptions 1 and 2 all closed-loop transfer functions are exp. stable.

We propose now to study the frequency responses $H_{yu_1}(j\omega)$ and $H_{yd_0}(j\omega)$ (i.e., of the closed-loop output due to input and disturbance at the output of the plant) under "large loop gain." From Fig. 1 one has

7 $$H_{yu_1} = PC(I + FPC)^{-1} = (I + PCF)^{-1}PC$$

8 $$H_{yd_0} = (I + PCF)^{-1},$$

9 $$H_{yu_1} - F^{-1} = -(I + PCF)^{-1}F^{-1},$$

where in (9) we have assumed that F is invertible. We shall denote by $\|P\|$, $\sigma_{max}[P]$, $\sigma_{min}[P]$ the ℓ_2-induced norm and the maximum and minimum singular value of the complex matrix transfer function P at $j\omega$, $\omega \in \mathbb{R}$. For feedback system Σ of Fig. 1, PCF, $(I + PCF)$, $(I + PCF)^{-1}$ are usually called the loop-, return difference- and sensitivity transfer functions.

Recall Theorem 1.2.57. Taking norms on both sides of (8) and (9), we note that if the sensitivity is small over Ω, equiv.

10 $$\|(I + PCF)^{-1}\| = \sigma_{max}[(I + PCF)^{-1}] << 1 \quad \text{over } \Omega,$$

where Ω is a bounded frequency interval (as $|\omega| \to \infty, (I + PCF)^{-1} \to I$), then

11 $$\|H_{yd_0}\| << 1 \quad \text{over } \Omega,$$

12 $$\|H_{yu_1} - F^{-1}\| << \|F^{-1}\| \quad \text{over } \Omega,$$

i.e., over Ω the closed-loop output y can be made insensitive to disturbances at the output of the plant and the I/O map: $u_1 \mapsto y$ is approximately equal to F^{-1}; i.e., we get response shaping over Ω.

Let us tie up (10) with "large loop gain."

16 Definition. We say that the m.i.m.o. feedback system Σ of Fig. 1 has a large loop gain over Ω iff

17 $$\sigma_{min}[PCF] >> 1 \quad \text{over } \Omega.$$ ■

18 Comment. In view of the s.v.d. of PCF at $j\omega$, condition (17) requires the size of the action of $PCF(j\omega)$ to be large in all directions $\forall\omega \in \Omega$.

The connection with (10) is

19 Lemma. Given the feedback system Σ of Fig. 1. U.t.c.

20 $$\|(I + PCF)^{-1}\| = \sigma_{max}[(I + PCF)^{-1}] << 1 \quad \text{over } \Omega$$

$\Leftrightarrow$

21 $$\sigma_{min}[I + PCF] >> 1 \quad \text{over } \Omega$$

$\Leftrightarrow$

17 $$\sigma_{min}[PCF] >> 1 \quad \text{over } \Omega.$$

22 Comment. Small sensitivity over Ω means a large return difference and large loop gain over Ω.

23 Proof of Lemma 19. Use ℓ_2-norms and Theorem 1.2.57.

(20) ⇔ (21): $\sigma_{max}[(I + PCF)^{-1}] = \|(I + PCF)^{-1}\| = (\sigma_{min}[I + PCF])^{-1}$.

(21) ⇔ (17): at $\omega \in \Omega$, $\forall x \in \mathbb{C}^{n_o}$ s.t. $\|x\| = 1$

$$\|PCFx\| - \|x\| \leq \|(I + PCF)x\| \leq \|PCFx\| + \|x\|.$$

Taking successive minima over $x \in \mathbb{C}^{n_o}$ with $\|x\| = 1$,

$$\sigma_{min}[PCF] - 1 \leq \sigma_{min}[I + PCF] \leq \sigma_{min}[PCF] + 1.$$

Hence

$$\sigma_{min}[PCF] >> 1 \Rightarrow \sigma_{min}[I + PCF] >> 1 \Rightarrow \sigma_{min}[PCF] >> 1.$$ ■

Hence by (10)-(12) and Lemma 19 one has

24 Theorem. Given feedback system Σ of Fig. 1. U.t.c.

17 $\sigma_{min}[PCF] >> 1$ over Ω

⇒

11 $\|H_{yd_0}\| << 1$ over Ω

12 $\|H_{yu_1} - F^{-1}\| << \|F^{-1}\|$ over Ω.

25 Comment. If the loop gain is large over Ω, the closed-loop output y can be made insensitive to disturbances applied at the plant output and the I/O-map is approximately equal to F^{-1}: for F = I we get approximate decoupling over frequency band Ω.

We study next the sensitivity of the closed-loop response w.r.t. plant variations.

28 Assumption 3. Considering Fig. 1, the nominal plant $P \in \mathbb{R}_{p,o}(s)^{n_o \times n_i}$ is replaced by a perturbed plant $\tilde{P} \in \mathbb{R}_{p,o}(s)^{n_o \times n_i}$ such that (a) the nominal feedback system Σ is replaced by a perturbed feedback system $\tilde{\Sigma}$ and (b) Assumptions 1 and 2 are still valid. ■

For the feedback configuration of Fig. 1 a plant variation

29 $$\Delta P = \tilde{P} - P$$

generates a variation of the closed-loop I/O map given by

30 $$\Delta H_{yu_1} = \tilde{H}_{yu_1} - H_{yu_1}.$$

By equation (7), the I/O map H_{yu_1} can also be obtained as the I/O map of the nominally equivalent open-loop system [Cru.1] shown in Fig. 2. For this open-loop configuration a plant variation (29) generates

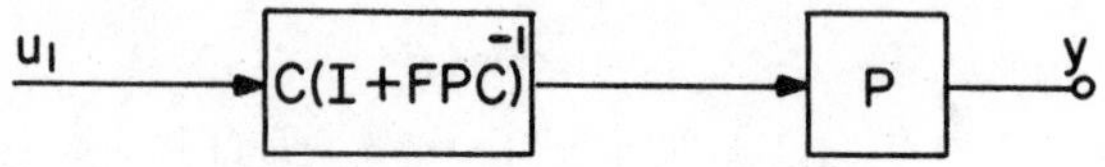

Fig. 2. The nominally equivalent open-loop system.

an I/O map variation

31 $$\Delta H^0_{yu_1} = \Delta P \cdot C(I + FPC)^{-1}.$$

Using (7) and (29)-(31) ΔH_{yu_1}, the variation of the closed-loop I/O map, is given in terms of $\Delta H^0_{yu_1}$, the variation in terms of the nominally equivalent open-loop system, by

32 $$\begin{aligned}\Delta H_{yu_1} &= (I + \tilde{P}CF)^{-1}\tilde{P}C - PC(I + FPC)^{-1} \\ &= (I + \tilde{P}CF)^{-1}\Delta PC(I + FPC)^{-1} = (I + \tilde{P}CF)^{-1}\Delta H^0_{yu_1}.\end{aligned}$$

We now have

35 Theorem. Consider the feedback system of Fig. 1 and the nominally equivalent open-loop system of Fig. 2 subject to plant variations (29) s.t. Assumption 3 holds.

U.t.c.

The closed-loop and nominally equivalent open-loop variations in the input-output map (30) and (31) are related by

36 $$\Delta H_{yu_1} = (I + \tilde{P}CF)^{-1}\Delta H^0_{yu_1}.$$

Therefore, if

37 $$\sigma_{min}[\tilde{P}CF] >> 1 \qquad \text{over } \Omega,$$

then

38 $$\|\Delta H_{yu_1}\| << \|\Delta H^0_{yu_1}\| \qquad \text{over } \Omega.$$ ■

39 <u>Comment</u>. <u>If the loop gain of the perturbed system is large over Ω, then the closed-loop I/O map variation is much smaller than that of the nominally equivalent open-loop system</u>.

42 <u>Proof of Theorem 35</u>. (36) has been obtained in (32). Taking norms at $\omega \in \Omega$, (36) gives

$$\|\Delta H_{yu_1}\| \leq \|(I + \tilde{P}CF)^{-1}\| \|\Delta H^0_{yu_1}\|.$$

Therefore, (37) implies (38) by Lemma 19. ■

The following Theorem compares "percentage changes" for the case of a <u>square</u> invertible plant [Saf.1].

43 <u>Theorem</u>. Consider the feedback system of Fig. 1 with $n_o = n_i$. Consider the variation of the closed-loop input-output map (30) generated by a plant variation (29) s.t. Assumption 3 holds.
U.t.c.
(a)

44 $$\Delta H_{yu_1} \cdot H^{-1}_{yu_1} = (I + \tilde{P}CF)^{-1}(\Delta P \cdot P^{-1}).$$

Consequently, if

45 $$\sigma_{min}[\tilde{P}CF] >> 1 \qquad \text{over } \Omega,$$

then

46 $$\|\Delta H_{yu_1} \cdot H^{-1}_{yu_1}\| << \|\Delta P \cdot P^{-1}\| \qquad \text{over } \Omega.$$

(b)

47 $$\Delta H_{yu_1} \cdot \tilde{H}^{-1}_{yu_1} = (I + PCF)^{-1}(\Delta P \cdot \tilde{P}^{-1}).$$

Consequently, if

17 $$\sigma_{min}[PCF] >> 1 \qquad \text{over } \Omega,$$

then

48 $$\|\Delta H_{yu_1} \cdot \tilde{H}^{-1}_{yu_1}\| << \|\Delta P \cdot \tilde{P}^{-1}\| \qquad \text{over } \Omega,$$ ✖

49 Comment. Formulas (44) and (47) relate relative changes in the closed-loop I/O map to relative change in the plant. These are w.r.t. nominal values in (44) and w.r.t. perturbed values in (47) while the reverse takes place for the sensitivities. The results are similar: for a large loop gain over Ω, the I/O map is made insensitive to plant variations.

50 Proof of Theorem 43. By (32) and (7) one obtains (44); (46) is obtained taking norms and Lemma 19.

51 Exercise. Prove assertions (47) and (48).

52 Conclusion. For the m.i.m.o. feedback system of Fig. 1 a large loop gain ensures

(1) that the closed-loop I/O map is approximately F^{-1} (roughly independent of C and P);

(2) desensitization of the closed-loop system output y due to disturbances applied at the output of the plant;

(3) densensitization of the closed-loop I/O map due to plant variations. ✖

53 Comment. Since F determines the I/O map over Ω (see (12)), it is the precompensator C that has to be made large in order to obtain a large loop gain PCF. It is well known that, in most cases, if the loop gain is made too large, the system becomes unstable. Also, if C becomes large, it will amplify any noise present at the input summing node and may cause plant saturation.

54 Comment. For many uses of singular values, see IEEE Trans. AC-26, No. 1, Feb. 1981. Some aspects of the nonlinear theory of large loop gain may be found in [Des.3], [Saf.1], and [Cru.1].

Chapter 2. Matrix Fraction Description of Transfer Functions

2.1. Introduction

We all know the state space description of linear lumped time-invariant systems:

1a $$\dot{x}(t) = Ax(t) + Bu(t)$$

$$t \geq 0$$

1b $$y(t) = Cx(t) + Du(t)$$

where A, B, C, D are constant matrices and $x(\cdot)$, $y(\cdot)$ and $u(\cdot)$ are, respectively, the state, the output, and the input of the system. With $p := d/dt$, these equations read

2a $$(pI-A)x(t) = Bu(t)$$

2b $$y(t) = Cx(t) + Du(t).$$

We may view these equations as specified by a quadruple of polynomial matrices $[pI - A, B, C, D]$. More generally we may consider a quadruple of polynomial matrices: $[D, N_\ell, N_r, K]$ and specify the system by

3a $$D(p)\xi(t) = N_\ell(p)u(t)$$

$$t \geq 0$$

3b $$y(t) = N_r(p)\xi(t) + K(p)u(t)$$

This is called a polynomial matrix system description (PMD) [Ros.1,Kai.1]. (For a precise definition, see Sec. 3.2.1.).

The PMD described by (3) has the transfer function

4 $$H = N_r D^{-1} N_\ell + K \in \mathbb{R}(s)^{n_o \times n_i}$$

In many physical problems, the linearized equations lead to expressions of the form

5 $$D_\ell(p)\xi(t) = N_\ell(p)u(t) \qquad y(t) = \xi(t),$$

with the transfer function

6 $$H = D_\ell^{-1}N_\ell \in \mathbb{R}(s)^{n_o\times n_i}.$$

In circuit theory, for example, the tableau equations lead to equations of the form

7 $$D_r(p)\xi(t) = u(t)$$
$$y(t) = N_r(p)\xi(t) + K(p)u(t)$$

with transfer function

8 $$H = N_rD_r^{-1} + K = (N_r + KD_r)D_r^{-1}$$

Equation (6) gives a left matrix fraction of H and (8) gives a right matrix fraction of H.

It is for this reason that we study in this chapter polynomial matrix fractions of rational matrices. Section 2 develops the algebraic tools for studying polynomial matrices (the Euclidean ring of polynomials, modules of polynomial vectors, $\cdots$). Section 3 studies polynomial matrices: first, algebraically, (coprimeness, elementary operations, greatest common divisors, $\cdots$), and then dynamically (the differential equation $D(p)\xi(t) = \theta$, $\cdots$). Section 4 studies polynomial matrix fractions: first algebraically, (coprime and proper fractions, division $\cdots$) and then the related system theoretic properties (poles and zeros, dynamical interpretations, $\cdots$). Section 5 sketches how a matrix fraction transfer function can be realized as a state space system.

2.2. Polynomials, Euclidean Rings, and Modules

In this section we describe some algebraic structures and facts related to polynomials and polynomial matrices. The reader is assumed familiar with the algebraic appendices A and B.

1 Definition. A Euclidean ring $(R, +, \cdot, 0, 1)$ is a commutative entire ring, (A.10), upon which there is defined a gauge [Sig. 1,p.132], i.e., a function

$$\gamma : R\backslash\{0\} \to \mathbb{N} : a \mapsto \gamma(a)$$

s.t. the following axioms hold:

ER1. $\forall a, b \in R\backslash\{0\}$ $\gamma(a) \leq \gamma(ab)$

ER2. If $a \in R$, $b \in R\backslash\{0\}$, then $\exists$ elements q and r in R s.t. $a = bq + r$, where $r = 0$ or $\gamma(r) < \gamma(b)$. ∎

2 Comment. Roughly speaking, a Euclidean ring is an entire ring in which a division operation is defined and which delivers a quotient q and a remainder r: ER2 is called the Euclidean division property.

3 Fact. $(\mathbb{R}[s], +, \cdot, 0, 1)$ is a Euclidean ring under pointwise addition and multiplication with $\gamma(a) := \partial a$, or equiv. the gauge of any polynomial is its degree

4 Indication. Fact 3 is based on the properties of the degree of a polynomial and the Euclidean Division Theorem. Let $a, b \in \mathbb{R}[s]$ with $b \neq 0$; then $\exists!$ q, r in $\mathbb{R}[s]$ s.t.

$$a = bq + r \quad \text{with} \quad r = 0 \quad \text{or} \quad \partial r < \partial b.$$

Note that (a) if $\partial a < \partial b$, then $q = 0$ and $r = a$, and (b) by convention $\partial 0 := -\infty$.

5 Exercise. Show that the division of $a(s) := 4s^2 + 3s + 1$ by $b(s) := 2s + 1$ gives $q(s) = 2s + .5$ and $r(s) = .5$.

In control theory the following are also important Euclidean rings.

6 Fact [Hun.1] [Cal.1]. Consider $R(0)$ the ring of exp. stable transfer functions:

$$R(0) := \{f \in \mathbb{R}_p(s) : f \text{ is analytic in } \mathbb{C}_+\}.$$

Then $(R(0), +, \cdot, 0, 1)$ is a Euclidean ring under pointwise addition and pointwise multiplication where the gauge of any element $f \in R(0)$ is the number of zeros of f in $\mathbb{C}_+$ and at ∞ (the number of zeros of f at ∞ is the difference in degree between the denominator and the numerator of f).

7 Fact. Any field $(\mathbb{F}, +, \cdot, 0, 1)$ is a Euclidean ring with a gauge γ s.t. $\gamma(a)$ = nonzero constant, $\forall a \in \mathbb{F}\backslash\{0\}$.

8 Comment. An arbitrary Euclidean ring is not a field because every nonzero element has not necessarily a multiplicative inverse (for example, $\mathbb{R}[s]$).

9 Fact. $a \in \mathbb{R}[s]$ has an inverse in $\mathbb{R}[s]$, equiv. $a^{-1} = 1/a \in \mathbb{R}[s]$, iff $a(s) \equiv k$, where k is a nonzero constant.

10 Fact. $f \in R(0)$ has an inverse in $R(0)$, equiv. $f^{-1} = 1/f \in R(0)$, iff $f(s) \neq 0\ \forall s \in \mathbb{C}_+$ and at ∞.

Every engineer has encountered the vector spaces $\mathbb{R}^n$, $\mathbb{R}(s)^n$ over the fields $\mathbb{R}$, $\mathbb{R}(s)$ $\cdots$. The set of polynomial n-vectors denoted by $\mathbb{R}[s]^n$ is closed under addition and multiplication by elements of the ring $\mathbb{R}[s]$, which is not a field $\cdots$.

15 Definition. Let R be a commutative ring. A module M over R, denoted by $(M, R, +, \cdot, \theta_M)$, is a set M together with a commutative ring R such that

M1. $(M, +, \theta_M)$ is an additive commutative group, equiv.

$\exists$ a binary operation + called addition and given by

$$+ : M \times M \to M : (m_1, m_2) \mapsto m_1 + m_2$$

s.t.

(a) Addition is associative and commutative

(b) $\exists!\ \theta_M \in M$ s.t.

$$\forall m \in M,\ m + \theta_M = \theta_M + m = m$$

(c) $\forall m \in M\ \ \exists!\ -m \in M$ s.t.

$$m + (-m) = (-m) + m = \theta_M.$$

(Notation: $\forall m_1, m_2 \in M$ we write $m_1 - m_2$ to denote $m_1 + (-m_2)$.)

M2. The module is closed under multiplication $\cdot$ by scalars, equiv.

$$\cdot : R \times M \to M : (r, m) \mapsto rm,$$

where $\forall m \in M$

$$1m = m, \quad 0m = \theta_M.$$

M3. Addition and multiplication by scalars are related by distributive laws, viz.,

$$\forall m \in M,\ \forall r_1, r_2 \in R,\ (r_1 + r_2)m = r_1 m + r_2 m;$$

$$\forall m_1, m_2 \in M,\ \forall r \in R,\ r(m_1 + m_2) = rm_1 + rm_2. \qquad ◼$$

The definitions and facts below are also needed to explain $\mathbb{R}[s]^n$.

16 Definitions. Let $(M, R, +, \cdot, \theta_M)$ be a module.

(a) $S \subset M$ is a submodule if $(S, R, +, \cdot, \theta_M)$ is a module under the same operations as for M.

(b) A submodule generated by a subset $S \subset M$ is the intersection of all submodules containing S, or equiv. the smallest submodule containing S.

17 Definitions. Let M_1 and M_2 be two modules over the same ring R. The product module $M_1 \times M_2$ over R is the module $(M_1 \times M_2, R, +, \cdot, (\theta_{M_1}, \theta_{M_2}))$ using componentwise addition and multiplication by scalars.

18 Fact. $(\mathbb{R}[s]^n, \mathbb{R}[s], +, \cdot, \theta_n)$ is a module: it is the n-fold product module of $(\mathbb{R}[s], \mathbb{R}[s], +, \cdot, 0)$: a module under pointwise addition and multiplication.

20 Definitions. Let (M, R) denote a module M over the ring R. Let I be a finite index set. (a) We say that $(m_i)_{i \in I} \subset M$ is a linearly dependent family of (M, R) iff

$\exists$ scalars $r_i \in R$, $i \in I$, not all zero, such that

$$\sum_{i \in I} r_i m_i = \theta_M .$$

(b) We say that $(m_i)_{i \in I}$ is a basis of (M, R) iff (1) $(m_i)_{i \in I}$ is a linearly independent family and (2) the smallest submodule generated by $(m_i)_{i \in I}$ is M. (Note that not all R-modules have a basis because R is not a field [Sig.1].)

(c) A module M which has a basis is called a free module; the dimension of a free module M is by definition the cardinality of any basis of M.

21 Fact. $(\mathbb{R}[s]^n, \mathbb{R}[s], +, \cdot, \theta_n)$ is a free module of dimension n having the basis

$$e_i = (0, \cdots, 0, 1, 0, \cdots, 0)^T \in \mathbb{R}[s]^n \quad i \in \underline{n}.$$
$$\uparrow$$
$$\text{ith-component}$$

22 Important Comment. A vector space V over the field $\mathbb{F}$, i.e., $(V, \mathbb{F}, +, \cdot, \theta_V)$, is similarly defined as a module $(M, R, +, \cdot, \theta_M)$, and has similar definitions for "subspace," "subspace generated by $S \subset V$," "product space," "linear dependence," "basis," and "dimension." A vector space always has a basis [Sig.1].

23 Exercise. Show that
(a) A vector space $(V, \mathbb{F})$ is a module (M, R);
(b) $(\mathbb{R}[s]^n, \mathbb{R}(s), +, \cdot, \theta_n)$ is not a module;
(c) $(R(0)^n, R(0), +, \cdot, \theta_n)$ is a free module of dimension n; is a subset of the n-dimensional vector space $(\mathbb{R}(s)^n, \mathbb{R}(s), +, \cdot, \theta_n)$; $(R(0)^n, R(0))$ and $(\mathbb{R}(s)^n, \mathbb{R}(s))$ have a common basis.
(d) $(\mathbb{R}[s]^2, \mathbb{R}[s])$ is a submodule of $(\mathbb{R}[s]^3, \mathbb{R}[s])$; is the submoaule generated by $S = \{(1, 0, 0)^T; (0, 1, 0)^T\}$.
(e) Show that $S := \{(s, 0), (0, 1)\}$ is not a basis for $(\mathbb{R}[s]^2, \mathbb{R}[s])$, but constitutes a linearly independent family. Determine the submodule generated by S.

Since the ring of polynomials $\mathbb{R}[s]$ is contained in the field of rational functions $\mathbb{R}(s)$, we have the following important fact.

26 Theorem. Consider the module $(\mathbb{R}[s]^n, \mathbb{R}[s], +, \cdot, \theta_n)$. Then,
(a)

27 $$(\mathbb{R}[s]^n, \mathbb{R}[s], +, \cdot, \theta_n) \subset (\mathbb{R}(s)^n, \mathbb{R}(s), +, \cdot, \theta_n)$$

We say that the $\mathbb{R}[s]$-module of polynomial n-vectors is embedded in the $\mathbb{R}(s)$-vector space of rational function n-vectors;
(b) with $(m_i)_{i\in I}$ a finite family of elements of the module $(\mathbb{R}[s]^n, \mathbb{R}[s])$

28 $(m_i)_{i\in I}$ is linearly dependent over $\mathbb{R}[s]$

$\Leftrightarrow$

29 $(m_i)_{i\in I}$ is linearly dependent over $\mathbb{R}(s)$

Comment. (b) means also that $(m_i)_{i\in I}$ is linearly independent (ℓ.i.) over $\mathbb{R}[s]$ iff this holds over the field $\mathbb{R}(s)$; consequently, the $\mathbb{R}[s]$-submodule generated by $(m_i)_{i\in I}$ inherits the dimension of the $\mathbb{R}(s)$-subspace generated by

$(m_i)_{i\in I}$: a question handled by classical vector space methods.

Proof of Theorem 26. (a) is obvious.

(b) $\Rightarrow$: is also obvious since $\mathbb{R}[s] \subset \mathbb{R}(s)$.

$\Leftarrow$: By assumption $\exists$ scalars $r_i \in \mathbb{R}(s)$, $i \in I$, not all zero s.t. $\sum_{i\in I} r_i m_i = \theta_r$. Let d be a least common denominator of the r_i's; then $dr_i \in \mathbb{R}[s]$, $\forall i \in I$, and the dr_i's are not all zero s.t. $\sum_{i\in I} dr_i m_i = \theta_n$. ■

A polynomial matrix $M \in \mathbb{R}[s]^{m\times n}$ is the representation of a linear map sending the module $(\mathbb{R}[s]^n, \mathbb{R}[s])$ into the module $(\mathbb{R}[s]^m, \mathbb{R}[s])$. We say that M is an $m \times n$ matrix over the ring $\mathbb{R}[s]$. Note that matrices over a ring do not have the same properties as matrices over a field: in particular, invertibility, as shown presently.

33 Theorem. Consider $M \in \mathbb{R}[s]^{n\times n}$ as a linear map over $\mathbb{R}[s]$-modules, i.e., $M : (\mathbb{R}[s]^n, \mathbb{R}[s]) \to (\mathbb{R}[s]^n, \mathbb{R}[s]) : x \mapsto Mx$

U.t.c.

(a)

34 $\det M \neq 0 \Leftrightarrow$ the columns of M are linearly independent over $\mathbb{R}[s]$

$\Leftrightarrow Mx = \theta_r \Rightarrow x = \theta_n$

$\Leftrightarrow$ linear map M is injective (1 to 1).

(b)

35 $\det M = k$, a nonzero constant, $\Leftrightarrow M^{-1} \in \mathbb{R}[s]^{n\times n}$,

$\Leftrightarrow$ {linear map M is injective and surjective (onto).

36 Comments. (a) Condition (34) is necessary and sufficient for M to be invertible over the field $\mathbb{R}(s)$ but not over the ring $\mathbb{R}[s]$; note (35). Consider $M = \text{diag}[s + 1, s + 2] \in \mathbb{R}[s]^{2\times 2}$, then $M = \text{diag}[(s + 1)^{-1}, (s + 2)^{-1}] \in \mathbb{R}(s)^{2\times 2}$ but $\notin \mathbb{R}[s]^{2\times 2}$: the map $M : \mathbb{R}[s]^2 \to \mathbb{R}[s]^2$ is not surjective.

(b) Similar results hold for M^T, i.e., the transpose of M.

37 Definition. A polynomial matrix $M \in \mathbb{R}[s]^{n\times n}$ satisfying (34), resp. (35) is called nonsingular, resp. unimodular or invertible (nonsingular polynomial matrices are not necessarily invertible although their rational inverse exists!)

38 Proof of Theorem 33. (a) According to Theorem 26 one has: the columns of M are ℓ.i. over $\mathbb{R}[s]$ iff they are ℓ.i. over $\mathbb{R}(s)$. Now $M \in \mathbb{R}(s)^{n\times n}$: a matrix over a field. So the latter condition is equivalent to det $M \neq 0_{\mathbb{R}(s)} = 0_{\mathbb{R}[s]}$.
(b) According to Fact B.7, $M^{-1} \in \mathbb{R}[s]^{n\times n}$ iff $(\det M)^{-1} \in \mathbb{R}[s]$: the latter condition is equivalent to $\det M \equiv k$ a nonzero constant by Fact 9. ∎

2.3. Polynomial Matrices

In this section we consider polynomial matrices $M \in \mathbb{R}[s]^{m\times n}$. All algebraic definitions and facts below hold for matrices $M \in R^{m\times n}$, where R is a Euclidean ring modulo some small modifications. Proofs concerning the rank are specific for polynomial matrices.

2.3.1. Divisors, Coprimeness, Rank

We first discuss factoring a polynomial matrix. Avoid reading parallel definitions in a first reading.

1 Definitions. (a) Let A, B, $C \in E(\mathbb{R}[s])$ with $A = BC$. Then C is said to be a right divisor (r.d.) of A and A is a left multiple (ℓ.m.) of C; similarly, B is a left divisor (ℓ.d.) of A and A is a right multiple (r.m.) of B.
(b) Similarly using matrices in $(\mathbb{R}[s])$, let $A = A_1R = LA_2$ and $B = B_1R = LB_2$. Then R (L) is said to be a common right divisor (c.r.d.) (common left divisor (c.ℓ.d.)) of A and B. If in addition R (L) is a ℓ.m. (r.m.) of every c.r.d. (c.ℓ.d.) of A and B, then R (L) is said to be a greatest common right divisor (g.c.r.d.) (greatest common left divisor (g.c.ℓ.d.)) of A and B.
(c) Two matrices A, $B \in E(\mathbb{R}[s])$ with the same number of columns (rows) are said to be right-coprime (r.c.) (left-coprime (ℓ.c.)) iff they have a g.c.r.d. (g.c.ℓ.d.) which is unimodular.

2 Comment. A better terminology for divisor is factor: what is really meant is divisor without remainder.

3 Exercise. Show that if R is a g.c.r.d. of A and B as in Definition 1, then LR, with L a unimodular polynomial matrix, is also a g.c.r.d. (i.e., g.c.r.d.'s are not unique).

We next discuss the rank of a polynomial matrix.

5 Definitions. Let $M \in \mathbb{R}[s]^{m\times n}$ and let r be an integer s.t. $0 \leq r \leq \min(m,n)$.
(a) We say that M has normal (determinantal) rank r iff ∃ at least one $r \times r$ minor which is not the zero polynomial and $\forall s > r$ every $s \times s$ minor is the zero polynomial. We say also that M has rank r over $\mathbb{R}[s]$. This is denoted by

6 $$\text{rk } M = r.$$

(b) We say that M has local rank r at $s \in \mathbb{C}$ iff the matrix $M(s) \in \mathbb{C}^{m\times n}$ has rank r. This is denoted by

7 $$\text{rk}[M(s)] = r \quad \text{or} \quad \text{rk}M = r \text{ at } s.$$ ∎

8 Definition. For $M \in \mathbb{R}(s)^{m\times n}$ the normal rank is defined as rank over the field $\mathbb{R}(s)$. For any $s \in \mathbb{C}$ not a pole of M the local rank of M at s is the rank of $M(s) \in \mathbb{C}^{m\times n}$.

10 Theorem. Let $M \in \mathbb{R}[s]^{m\times n}$; then

6 $\text{rk}M = r$

⇔

11 r = maximum # of ℓ.i. columns of M in $(\mathbb{R}[s]^m, \mathbb{R}[s])$(=: normal column rank),

⇔

12 r = maximum # of ℓ.i. rows of M in $(\mathbb{R}[s]^n, \mathbb{R}[s])$(=: normal row rank).

13 Comment. By Theorem 10 for any polynomial matrix $M \in \mathbb{R}[s]^{m\times n}$ the normal determinantal rank = the normal column rank = the normal row rank: this common rank is called the normal rank of M.

14 Proof of Theorem 10. Observe that $M \in \mathbb{R}[s]^{m\times n} \subset \mathbb{R}(s)^{m\times n}$. Now for matrices over a field the assertion of the comment is true. The theorem follows therefore because (1) the determinantal rank over $\mathbb{R}[s]$ equals the determinantal rank over $\mathbb{R}(s)$ and (2) by Theorem 2.2.26 the maximum # of ℓ.i. columns (rows) over $\mathbb{R}[s]$ equals the maximum # of ℓ.i. columns (rows) over $\mathbb{R}(s)$. ∎

From the proof of Theorem 10, one has also

15 Theorem. Let $M \in \mathbb{R}[s]^{m\times n}$; then the rank of M over $\mathbb{R}[s]$ = the rank of M over $\mathbb{R}(s)$, i.e., M has the same normal rank as a polynomial and as a rational matrix.

16 Comment. Note that by Theorem 15 linear independence over the field $\mathbb{R}(s)$ may be used to check the normal row or column rank of M as a polynomial matrix.

Another consequence of Theorem 10 is

17 Exercise. Let $M \in \mathbb{R}[s]^{m\times n}$. Show that $\text{rk}M = \text{rk}M^T$.

The exercises below relate normal rank, invertibility and local rank.

18 Exercise. Let $M \in \mathbb{R}[s]^{m\times n}$. Show that

$$\text{rk}M = r \Leftrightarrow \text{rk}[M(s)] = r \text{ except for at most a finite number of points } s \in \mathbb{C}.$$

(Hint: Nonzero minors have at most a finite number of zeros.) ∎

20 Exercise. Let $M \in \mathbb{R}[s]^{n\times n}$. Show that

(a)

$$M \text{ is nonsingular} \Leftrightarrow \text{rk}M = n$$

$$\Leftrightarrow \text{rk}[M(s)] = n \text{ except for at most a finite number of points } s \in \mathbb{C}.$$

(b)

$$M \text{ is unimodular} \Leftrightarrow \text{rk}[M(s)] = n \ \forall s \in \mathbb{C}.$$ ∎

22 Note. If $M \in \mathbb{R}[s]^{m\times n}$ and $\text{rk}M = m$, ($\text{rk}M = n$), then M is said to have full row rank (full column rank) over $\mathbb{R}[s]$.

2.3.2. Elementary Operations on Polynomial Matrices

In this section we list elementary operations and their properties.

1 Let $M \in \mathbb{R}[s]^{m\times n}$. Elementary row operations (e.r.o.'s) on M are of three kinds:

(1) Interchange two rows $\rho_i \leftrightarrows \rho_j$.
(2) Multiply a row by a nonzero <u>constant</u> k (i.e., an invertible element of $\mathbb{R}[s]$): $\rho_i \leftarrow k\rho_i$.
(3) For $j \neq i$ add to row i another row j multiplied by $r \in \mathbb{R}[s]$: $\rho_i \leftarrow \rho_i + r\rho_j$.

Note that e.r.o.'s are equivalent to premultiplying M by <u>left elementary matrices</u> (ℓ.e.m.'s) L; these are obtained from the unit matrix by performing the desired e.r.o. upon it. The ℓ.e.m.'s corresponding to the e.r.o.'s above are listed below; any entry not explicitly shown is zero; dots indicate entries which are 1; broken lines indicate affected rows and columns.

$$(1):\ L = \begin{array}{c} \\ \\ \\ \\ i \\ \\ \\ \\ j \\ \\ \\ \\ \end{array} \begin{bmatrix} 1 &&&&&&&&& \\ & \cdot &&&&&&&& \\ && \cdot &&&&&&& \\ &&& 1 &&&&&& \\ &&&& 0 &&& 1 && \\ &&&&& 1 &&&& \\ &&&&&& \cdot &&& \\ &&&&&&& 1 && \\ &&&& 1 &&&& 0 & \\ &&&&&&&&& 1 \\ &&&&&&&&& \cdot \\ &&&&&&&&& 1 \end{bmatrix} \in \mathbb{R}[s]^{m\times m},$$

(columns i and j marked above the matrix)

$$(2):\ L = \begin{array}{c} \\ \\ \\ \\ \\ i \\ \\ \\ \end{array} \begin{bmatrix} 1 &&&&&&& \\ & \cdot &&&&&& \\ && \cdot &&&&& \\ &&& \cdot &&&& \\ &&&& 1 &&& \\ &&&&& k && \\ &&&&&& 1 & \\ &&&&&&& 1 \end{bmatrix} \in \mathbb{R}[s]^{m\times m},$$

(column i marked above the matrix)

$$(3):\ L = \begin{array}{c} \\ \\ \\ i \\ \\ \\ j \\ \\ \\ \end{array} \begin{bmatrix} 1 &&&&&&& \\ & \cdot &&&&&& \\ && \cdot &&&&& \\ &&& 1 && r && \\ &&&& \cdot &&& \\ &&&&& \cdot && \\ &&&&& 1 && \\ &&&&&& \cdot & \\ &&&&&&& 1 \end{bmatrix} \in \mathbb{R}[s]^{m\times m}.$$

(columns i and j marked above the matrix)

2 <u>Elementary column operations</u> (e.c.o.'s) are similarly defined: replace "row" by "column" ($\rho_i \leftarrow \gamma_i$; $\rho_j \leftarrow \gamma_j$). E.c.o.'s are equivalent to

postmultiplying M by *right elementary matrices* (r.e.m.'s) R; these are obtained from the unit matrix by performing the desired e.c.o. upon it.

3 *Exercise*. Write down the three e.c.o.'s and their corresponding r.e.m.'s $R \in \mathbb{R}[s]^{n\times n}$.

4 *Definitions*. (a) A square polynomial matrix is said to be an *elementary matrix* (e.m.) iff it is a ℓ.e.m. or a r.e.m..
(b) An operation on a polynomial matrix is said to be an *elementary operation* (e.o.) iff it is an e.r.o. or an e.c.o..

Elementary operations are used to reduce polynomial matrices to forms that display wanted information; see Sec. 2.3.4 on standard forms. The following are important properties.

7 *Exercise*. Show that *each e.o. is invertible*, hence *each e.m. is unimodular* (equiv. invertible).
(Hint: $\rho_j \rightleftarrows \rho_i$; $\rho_i \leftarrow k^{-1}\rho_i$; ····; det L = k a nonzero constant ···.)

8 *Theorem* [Rank invariances]. Each elementary operation on a polynomial matrix leaves the normal rank unchanged.

Equivalently, let $M \in \mathbb{R}[s]^{m\times n}$ be a polynomial matrix and let $L \in \mathbb{R}[s]^{m\times m}$ and $R \in \mathbb{R}[s]^{n\times n}$ be an arbitrary left-, resp. right-, elementary matrix; then

$$\mathrm{rk}M = \mathrm{rk}LM, \quad \mathrm{rk}M = \mathrm{rk}MR.$$

9 *Comment*. This property is a key tool for the reduction to standard forms.

10 *Proof*. Note (a) $M \in \mathbb{R}(s)^{m\times n}$: a matrix over a field; (b) the e.o.'s listed above are also e.o.'s for matrices over the field $\mathbb{R}(s)$; (c) the latter operations leave the rank over $\mathbb{R}(s)$ unchanged; (d) by Theorem 2.3.2.15 the rank over $\mathbb{R}(s)$ is the rank over $\mathbb{R}[s]$. Hence the theorem. ∎

The last part of this section catches e.o.'s into the compact notion of *equivalence*. A forward reference is necessary.

13 *Theorem*. Let $M \in \mathbb{R}[s]^{m\times m}$.

M is unimodular

$\Leftrightarrow$

M is a finite product of elementary matrices.

Proof. $\Leftarrow$: follows by Exercise 7.
$\Rightarrow$: If M is unimodular, then it can be reduced by e.o.'s to a Smith form which is the unit matrix: see Exercise 2.3.4.30 below. ∎

14 Definitions. Let A and $B \in \mathbb{R}[s]^{m\times n}$; then A and B are said to be equivalent, left equivalent, right equivalent iff there exist unimodular matrices $L \in \mathbb{R}[s]^{m\times m}$ and $R \in \mathbb{R}[s]^{n\times n}$ s.t. resp. $A = LBR$, $A = LB$, $A = BR$. These relations are denoted by resp. $A \sim B$, $A \overset{\ell}{\sim} B$, $A \overset{r}{\sim} B$.

15 Exercise. Show that the relations defined in Definitions 14 are equivalence relations.
(Hint: $A \sim A$; $A \sim B \Rightarrow B \sim A$; $A \sim B$, $B \sim C \Rightarrow A \sim C$; $\cdots$.)

16 Exercise. With matrices A and B as in Definitions 14, show that A and B are equivalent, left equivalent, right equivalent iff A and B can be obtained from B and A by resp. e.o.'s, e.r.o.'s, e.c.o.'s.
(Hint: Use Theorem 13.)

17 Exercise. Show that the relations of equivalence, left equivalence, and right equivalence do not change the rank.
(Hint: Use Exercise 16 and Theorem 8.)

2.3.3. Elementary Operations and Differential Equations

Consider the differential equation

1 $$D(p)\xi(t) = \theta_\nu \quad \forall t \geq 0$$

where $p = d/dt$ is the differential operator, $D(\cdot) \in \mathbb{R}[p]^{\nu\times\nu}$ is a nonsingular polynomial matrix in p. For example,

2 $$D(p) = D_2p^2 + D_1p + D_0 \qquad D_i \in \mathbb{R}^{\nu\times\nu} \text{ for } i = 0 \sim 2.$$

Note that (1) is a differential equation with constant coefficients; for such an equation it is well known that any solution $\xi(\cdot)$ with values in $\mathbb{R}^\nu$ is a sum of exponential polynomials in t of the form

$$\sum_{k=1}^{m} (a_k t^{k-1})\exp(\lambda t), \quad \forall t > 0-,$$

where $a_k \in \mathbb{C}^\nu$ $\forall k \in \underline{m}$, $\lambda \in \mathbb{C}$ and $\det D(\lambda) = 0$ with the restriction that iff $\lambda \in \mathbb{C}\backslash\mathbb{R}$, then the solution has a complex conjugate companion term given by

$$\sum_{k=1}^{m} (\bar{a}_k t^{k-1})\exp(\bar{\lambda} t), \quad \forall t > 0-.$$

As a consequence, solutions of (1) are <u>infinitely differentiable on $(0-,\infty)$</u>: in particular, $t \to \xi(t)$ and all its derivatives are continuous at $t = 0$.

The reason we emphasize the behavior of $\xi(\cdot)$ at $t = 0$ is that when the input $u(\cdot)$ will be introduced, the solution of $D(p)\xi(t) = N_\ell(p)u(t)$, may be discontinuous at $t = 0$: hence it will be important then to distinguish between $\xi(0-)$ and $\xi(0+)$.

Hence the following definition.

3 <u>Definition</u>. A <u>solution</u> of (1) is a function $\xi(\cdot)$: $(0-,\infty) \to \mathbb{R}^\nu$ s.t. $D(p)\xi(t) = \theta_\nu$ $\forall t \geq 0$. The set of solutions of (1) is denoted by X.

4 <u>Fact</u>. The set X of solutions of (1) is an <u>$\mathbb{R}$-vector space</u>, equiv. if $\xi_1(\cdot)$, $\xi_2(\cdot) \in X$ and $a_1, a_2 \in \mathbb{R}$, then $a_1\xi_1(\cdot) + a_2\xi(\cdot) \in X$.

5 <u>Exercise</u>. Prove Fact 4. (Hint: Use (2) and observe that real matrix- and differential operators are linear.)

The consequences of e.o.'s over $\mathbb{R}[p]$ on polynomial matrix $D(p)$ in (1) are now explained. We start with e.r.o.'s.

6 <u>Theorem</u>. Let $D(\cdot) \in \mathbb{R}[p]^{\nu\times\nu}$ be nonsingular and let $L(\cdot) \in \mathbb{R}[p]^{\nu\times\nu}$ be <u>unimodular</u> with

7 $$\bar{D}(p) = L(p)D(p).$$

Consider the differential equations

8 $$D(p)\xi(t) = \theta_\nu \quad t \geq 0; \quad \bar{D}(p)\bar{\xi}(t) = \theta_\nu \quad t \geq 0$$

with solution spaces X resp. $\bar{X}$.

U.t.c.

9 $$X = \bar{X}.$$

10 Comment. Performing e.r.o.'s on $D(p)$ does not affect the solution space of $D(p)\xi(t) = \theta, t \geq 0$.

Proof of Theorem 6. Since L is unimodular both $L(\cdot)$ and $L(\cdot)^{-1}$ are polynomial matrices, whence $\forall t \geq 0$

$$D(p)\xi(t) = \theta \Rightarrow L(p)D(p)\xi(t) = \bar{D}(p)\xi(t) = \theta;$$

$$\bar{D}(p)\bar{\xi}(t) = \theta \Rightarrow L(p)^{-1}\bar{D}(p)\bar{\xi}(t) = D(p)\bar{\xi}(t) = \theta.$$ ∎

18 Theorem. Let $D(\cdot) \in \mathbb{R}[p]^{\nu\times\nu}$ be nonsingular and $R(\cdot) \in \mathbb{R}[p]^{\nu\times\nu}$ be unimodular with

19 $$\bar{D}(p) = D(p)R(p).$$

Consider the differential equations

20 $$D(p)\xi(t) = \theta_\nu \quad t \geq 0; \quad \bar{D}(p)\bar{\xi}(t) = \theta_\nu \quad t \geq 0$$

with solution spaces X resp. $\bar{X}$.
U.t.c. the map

21 $$T : \bar{\xi}(\cdot) \in \bar{X} \mapsto R(p)\bar{\xi}(\cdot) =: \xi(\cdot) \in X$$

is a linear bijection (i.e., an isomorphism) from $\bar{X}$ onto X.

22 Comments. E.c.o.'s on $D(p)$ result in a "change of variables" for the solutions of $D(p)\xi(t) = \theta$, $t \geq 0$. ∎

23 Proof of Theorem 18. Note that since $R(\cdot)$ is unimodular, $R(\cdot)$ and $R(\cdot)^{-1}$ are polynomial matrices. Hence

(a) $\forall t \geq 0$ with $\xi(\cdot) := R(p)\bar{\xi}(\cdot)$, where $\bar{\xi}(\cdot) \in \bar{X}$,

$$D(p)\xi(t) = D(p)R(p)\bar{\xi}(t) = \bar{D}(p)\bar{\xi}(t) = \theta.$$

Hence T defined by (21) maps $\bar{X}$ into X.

(b) $\forall t \geq 0$ with $\bar{\xi}(\cdot) = R(p)^{-1}\xi(\cdot)$, where $\xi(\cdot) \in X$,
$\bar{D}(p)\bar{\xi}(t) = \bar{D}(p)R(p)^{-1}\xi(t) = D(p)\xi(t) = \theta$.

Hence T defined by (21) is invertible.

The linearity of map (21) is left as an exercise. ∎

The last theorem of this section concerns itself with the effects of arbitrary e.o.'s, i.e., equivalence on $D(\cdot)$.

26 Theorem. Let $D(\cdot) \in \mathbb{R}[p]^{\nu\times\nu}$ be nonsingular and let $L(\cdot) \in \mathbb{R}[p]^{\nu\times\nu}$ and $R(\cdot) \in \mathbb{R}[p]^{\nu\times\nu}$ be unimodular matrices with

27 $$\tilde{D}(p) = L(p)D(p)R(p).$$

Consider the differential equations

28 $$D(p)\xi(t) = \theta_\nu, t \geq 0; \quad \tilde{D}(p)\bar{\xi}(t) = \theta_\nu, t \geq 0$$

with solution spaces X resp. $\bar{X}$.
U.t.c. the map

29 $$T : \bar{\xi}(\cdot) \in \bar{X} \mapsto R(p)\bar{\xi}(\cdot) =: \xi(\cdot) \in X$$

is a linear bijection (isomorphism) of $\bar{X}$ onto X. ∎

30 Exercise. Prove Theorem 26.
(Hint: Use Theorems 6 and 18.)

31 Comment. (a) In other words, $\bar{\xi}(\cdot)$ is a solution of (28^2) iff $\xi(\cdot)$ defined by (29) is a solution of (28^1).
(b) Isomorphic vector spaces have the same dimension; an isomorphism maps a basis onto a basis; etc.

2.3.4. Standard Forms: Hermite and Smith Forms

Every polynomial matrix $M \in \mathbb{R}[s]^{m\times n}$ can by e.o.'s be brought into a standard form: the most important are the Hermite form (upper or lower triangular) and the Smith form (diagonal).

1.L. Theorem [Hermite row form]. Let $M \in \mathbb{R}[s]^{m\times n}$. Then there exists a unimodular matrix $L \in \mathbb{R}[s]^{m\times m}$ (obtained by e.r.o.'s) such that

2 $LM = H =$

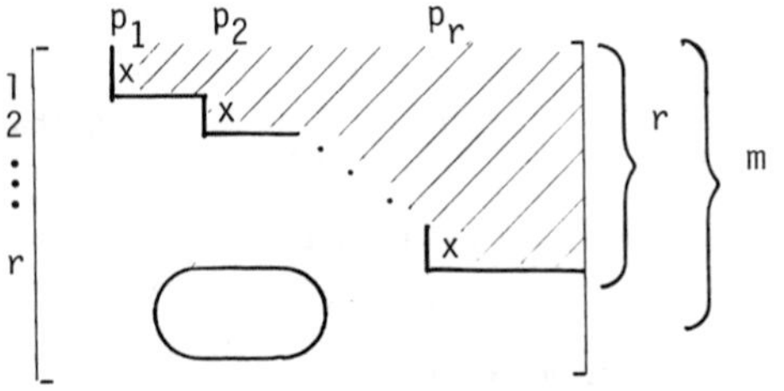

where H is called Hermite row form of M and has the following properties:

There exists an integer r, $0 \leq r \leq \min(m,n)$ with

3 $$\text{rk}H = \text{rk}M = r$$

s.t.

(a) $\forall i \in \underline{r}$, ρ_i has a leading nonzero monic polynomial h_{ip_i} called leading entry s.t.

$$1 \leq p_1 < p_2 < \cdots < p_r \leq n.$$

(b) $\forall i \in \underline{r}$,

if $h_{ip_i} = 1$, then $h_{jp_i} = 0 \quad \forall j < i$,

if $h_{ip_i} \neq 1$, then $\partial[h_{jp_i}] < \partial[h_{ip_i}] \quad \forall j < i$ s.t. $h_{jp_i} \neq 0$.

(c) $\forall \gamma_j$ s.t. $j < p_1$, γ_j is zero;

$\forall \gamma_j$ s.t. $p_i \leq j < p_{i+1}$ with $i \in \underline{r-1}$, then the last m-i entries of γ_j are zero;

$\forall \gamma_j$ s.t. $j \geq p_r$, the last m-r entries of γ_j are zero. ∎

4 Comment. Conditions (a)-(c) above describe the upper staircase form of H with r stairs. Note the similarity of the Hermite row form with the row echelon form for matrices $M \in \mathbb{R}^{m\times n}$ [Nob.1]. The difference is that we cannot annihilate every entry above a leading entry: since we are working with polynomials, the Euclid algorithm can only decrease the degree of such

nonzero entries below the degree of the leading entry. ∎

The Hermite row form is obtained by the following algorithm.

5.L. Algorithm [Reduction to Hermite row form]. For $i = 1, 2, \cdots$:

Step 1. Search for γ_{p_i}, the first column from the left that is nonzero below ρ_{i-1}.
If such column does not exist, or $i = m+1$, or $p_{i-1} = n$, STOP.

Step 2. Choose among the nonzero entries of γ_{p_i} below ρ_{i-1} an entry of smallest degree and by row permutation place it in position (i,p_i).

Step 3. Multiply ρ_i by a nonzero constant to make the entry in position (i,p_i) monic.

Step 4. Use the Euclid algorithm and addition of suitable polynomial multiples of ρ_i to reduce the nonzero entries in γ_{p_i} to their remainders after division by entry (i,p_i).

Step 5. If the remainders in γ_{p_i} below ρ_i are all zero, go to step 6. Else, repeat steps 2 to 4 until the remainders are all zero.

Step 6. If $i = 1$, skip step 6. Else, use the Euclid algorithm and addition of suitable polynomial multiples of ρ_i to reduce nonzero entries in γ_{p_i} above ρ_i to their remainders after division by entry (i,p_i). ∎

6 Exercise. Reduce to Hermite row form

$$M(s) = \begin{bmatrix} 0 & 0 & s^2 \\ s+1 & 0 & s^2 \\ s+2 & 0 & -s^2 \\ s+1 & 0 & -s^2 \end{bmatrix}.$$

7 Proof of Theorem 1.L. The proof is by construction and consists in justifying Algorithm 5.L.

A. Concerning step 5, note that the degrees of the nonzero remainders of the nonzero entries of γ_{ρ_i} below row i are smaller than the degree of entry (i,ρ_i). Hence by repeating steps 2 to 4 these degrees are steadily decreasing and zero remainders are ultimately obtained: step 6 will be reached after a finite number of sequences consisting of steps 2 to 4.

B. Since matrix M has a finite number of entries, the algorithm will stop in step 1 for a finite value of i =: r + 1 and the matrix will have a staircase form with exactly r stairs occurring in ρ_i, $\forall i \in \underline{r}$, as in (2).

C. Over the field $\mathbb{R}(s)$ the rank of H in (2) is clearly r. Since by Theorem 2.3.1.15 this is also the rank of H over $\mathbb{R}[s]$ and by Theorem 2.3.2.8 e.o.'s do not change the rank over $\mathbb{R}[s]$, we have

$$r = \text{rk}H = \text{rk}M.$$

Hence (3) holds.

D. It follows finally by construction that all properties of H listed under (a)-(c) of the theorem statement hold. ∎

8 Exercise. Read [e.g., Nob.1,pp.82-83] and observe that Algorithm 5.L is the Gauss-Jordan algorithm for obtaining the row-echelon form of $M \in \mathbb{R}^{m\times n}$ modulo modifications involving the use of the Euclid algorithm to reduce the degree of a polynomial and eventually making a polynomial zero.

9.L. Corollary. [Hermite row form of a full column rank matrix]. Let in Theorem 1.L $M \in \mathbb{R}[s]^{m\times n}$ have full column rank over $\mathbb{R}[s]$, equiv. rkM = n, whence $m \geq n$. Then there exists a unimodular matrix $L \in \mathbb{R}[s]^{m\times m}$ (obtained by e.r.o.'s) such that

$$LM = H = \overset{\displaystyle n}{\begin{bmatrix} R \\ \\ 0 \end{bmatrix}} \begin{matrix} n \\ \\ m-n \end{matrix}$$

with $R \in \mathbb{R}[s]^{n\times n}$ upper triangular and nonsingular; i.e., H, the Hermite row form of M, is upper triangular. Moreover, H is as in the statement of Theorem 1.L with

$$p_i = i \quad \forall i \in \underline{n}.$$

Proof. By the proof of Theorem 1.L, r = rkH is the number of stairs of the Hermite row form H given by (2). By assumption rkM = n, whence by (3) r = rkH = rkM = n. Hence matrix $H \in \mathbb{R}[s]^{m\times n}$ given by (2) must have exactly n stairs. The corollary follows. ∎

12 Convention. Up to now we obtained the Hermite row form of a polynomial matrix described by Theorem 1.L [Hermite row form], Algorithm 5.L [Reduction to Hermite row form] and Corollary 9.L [Hermite row form of a full column rank matrix]. Note that L stands for left operations (e.r.o.'s). Using e.c.o.'s instead of e.r.o.'s we have also Theorem 1.R [Hermite column form], Algorithm 5.R [Reduction to Hermite column form], and Corollary 9.R [Hermite column form of a full row rank matrix], where R stands for right operations (e.c.o.'s). This convention is used below. For reasons of completeness we give Theorem 1.R in detail.

1.R. Theorem [Hermite column form]. Let $M \in \mathbb{R}[s]^{m\times n}$. Then there exists a unimodular matrix $R \in \mathbb{R}[s]^{n\times n}$ (obtained by e.c.o.'s) such that

13 $\quad MR = H =$

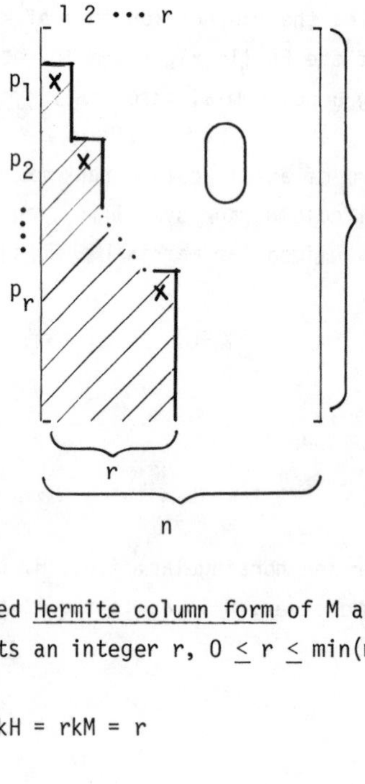

where H is called Hermite column form of M and has the following properties:

There exists an integer r, $0 \le r \le \min(m, n)$ with

14 $$\mathrm{rk}H = \mathrm{rk}M = r$$

s.t.

(a) $\forall j \in \underline{r}$, γ_j has a leading nonzero monic polynomial $h_{p_j j}$ called leading entry s.t.

$$1 \le p_1 < p_2 < \cdots < p_r \le m.$$

(b) $\forall j \in \underline{r}$,

$$\text{if } h_{p_j j} = 1, \text{ then } h_{p_j i} = 0 \quad \forall i < j,$$

$$\text{if } h_{p_j j} \neq 1, \text{ then } \partial[h_{p_j i}] < \partial[h_{p_j j}] \quad \forall i < j \text{ s.t. } h_{p_j i} \neq 0.$$

(c) $\forall \rho_i$ s.t. $i < p_1$, ρ_i is zero;

$\forall \rho_i$ s.t. $p_j \leq i < p_{j+1}$ with $j \in \underline{r-1}$, then the last $m - j$ elements of ρ_i are zero;

$\forall \rho_i$ s.t. $i \geq p_r$, the last $m - r$ elements of ρ_i are zero. ∎

15 *Exercise*. Write the statements of Algorithm 5.R [Reduction to Hermite column form] and Corollary 9.R [Hermite column form of a full row rank matrix]. (Hint: Use the statements of Algorithm 5.L and Corollary 9.L; replace e.r.o.'s by e.c.o.'s, "row" by "column," ⋯.)

Instead of using *only* e.r.o.'s (Hermite row form) or *only* e.c.o.'s (Hermite column form) we use now a combination of e.r.o.'s *and* e.c.o.'s to obtain a quasi-diagonal standard form.

18 *Theorem* [Smith form]. Let $M \in \mathbb{R}[s]^{m \times n}$. Then there exist *unimodular* matrices $L \in \mathbb{R}[s]^{m \times m}$ (obtained by e.r.o.'s) and $R \in \mathbb{R}[s]^{n \times n}$ (obtained by e.c.o.'s) such that

19 $$LMR = S = \left[\begin{array}{ccccc|c} \lambda_1 & & & & & \\ & \ddots & & 0 & & \\ & & \lambda_i & & & 0 \\ & 0 & & \ddots & & \\ & & & & \lambda_r & \\ \hline & & 0 & & & 0 \end{array}\right] \begin{array}{l} \}\, r \\ \\ \}\, m-r \end{array} \in \mathbb{R}[s]^{m \times n},$$

(column blocks: r, $n-r$)

where S is called the *Smith form of M* and

(a)

20 $$r = \text{rk}M = \text{rk}S \leq \min(m, n).$$

(b) The polynomials $\lambda_i \in \mathbb{R}[s]$, $i \in \underline{r}$, called *invariant polynomials of M*, are monic, *uniquely defined* by M, and satisfy the division property

21 $$\lambda_i | \lambda_{i+1} \qquad \forall i \in \underline{r-1}.$$

Moreover, consider the polynomials $\Delta_i \in \mathbb{R}[s]$, $i = 0 \sim r$, given by

22 $\Delta_0 \equiv 1$, Δ_i := the monic g.c.d. of all $i \times i$ minors of M.

These polynomials are called the determinantal divisors of M. They are related to the invariant polynomials of M by

23 $$\lambda_i = \Delta_i / \Delta_{i-1} \qquad i \in \underline{r}.$$ ∎

24 Comment. The top left block of the Smith form (19) is diagonal with unique invariant polynomials λ_i as diagonal entries: the Smith form is a unique standard form under equivalence.

25 Proof of Theorem 18. The proof is by construction and is carried out in four steps, A, B, C, and D.

26 A. Algorithm [Reduction to Smith form]

(1) By a succession of e.o.'s reduce matrix M into the form

27 $$\begin{array}{c} \\ 1\left\{\right. \\ \\ \end{array}\overbrace{}^{1}\left[\begin{array}{c|ccc} \lambda_1 & 0 & \cdots & 0 \\ \hline 0 & & & \\ \vdots & & M_1 & \\ 0 & & & \end{array}\right]$$

where λ_1 is monic and divides every element of M_1. This is done as follows:

Step 1. By a row and/or column permutation place a lowest degree nonzero entry in position (1, 1).

Step 2. If all entries of γ_1 and ρ_1 except for entry (1, 1) are zero, go to step 4.

Step 3. Find the remainders after division by entry (1, 1) of all entries of γ_1 and ρ_1 except for entry (1, 1).

If all these remainders are zero, then by the addition of suitable multiples of ρ_1 and γ_1 reduce the corresponding nonzero entries to zero and go to step 4.

Else consider a nonzero remainder of lowest degree and its corresponding nonzero entry. By the addition of a suitable multiple of ρ_1 or γ_1 replace that entry by its remainder and by a row or column permutation bring that remainder in position (1, 1). Next go to step 2.

Comment: The outcome of steps 2 and 3 is to reduce to zero every entry of γ_1 and ρ_1 except for entry (1, 1); note that the degree of entry (1, 1) steadily decreases when cycling occurs.

Step 4. If entry (1, 1) divides every entry of the matrix, STOP. Else there exists among the entries not in position (1, 1) a nonzero entry, say in γ_k, not divisible without remainder by entry (1, 1) and of lowest degree. Then add γ_k to γ_1 and repeat step 3. Keep on repeating until entry (1, 1) divides every entry of the matrix.

Comment: The degree of entry (1, 1) continues to decrease···. ✖

(2) Now operate similarly on M_1 in (27) to obtain

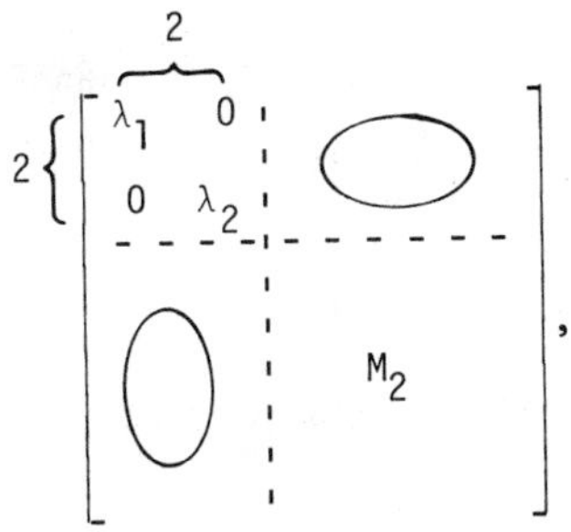

,

where λ_2 is monic and divides every element of M_2

⋮

(i) Operate similarly on M_{i-1} to obtain

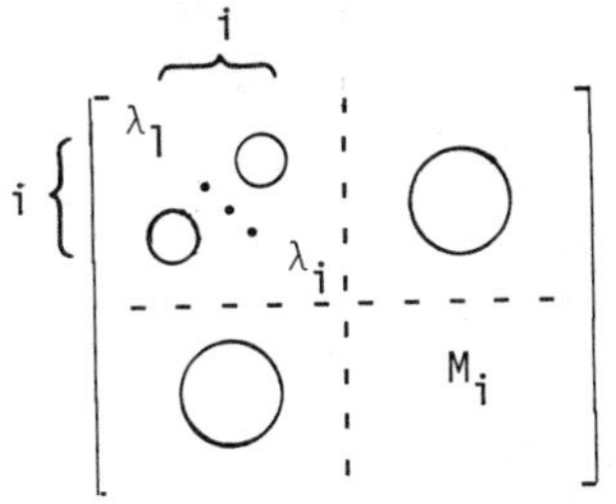

,

where λ_i is monic and divides every entry of M_i.

⋮

Do this until $i = \min(m, n)$ or $M_i = 0$, then STOP

End of Algo.

B. In the Smith Form Algorithm the sequence steps 1-4 is a finite procedure because the degree of the corner entry is steadily decreased. Moreover, the algorithm will stop when it runs out of submatrices M_i. Then, set $r :=$ the last value of i and the matrix will have form (19) with property (20).

C. In the Smith Form Algorithm, $\forall i \in \underline{r-1}$ there exist unimodular matrices L_i, R_i (obtained by e.o.'s) such that

$$L_i M_i R_i = \left[\begin{array}{c|ccc} \lambda_{i+1} & 0 & \cdots & 0 \\ \hline 0 & & & \\ \vdots & & M_{i+1} & \\ 0 & & & \end{array}\right]$$

and λ_i divides <u>every</u> entry of M_i. Obviously, λ_{i+1} is an $\mathbb{R}[s]$-linear combination of entries of M_i. Hence $\lambda_i | \lambda_{i+1}$: (21) holds.

D. Let

$\Delta_i^S :=$ the monic g.c.d. of all $i \times i$ minors of S,

$\Delta_i^M :=$ the monic g.c.d. of all $i \times i$ minors of M;

then by the equivalence relation (19)

$$LMR = S \quad \text{and} \quad M = L^{-1} S R^{-1},$$

where L, L^{-1}, R, and R^{-1} are polynomial square matrices. By the Cauchy-Binet formula [e.g., Kai. 1, pp. 649-650] it follows that every $i \times i$ minor of S is an $\mathbb{R}[s]$-linear combination of $i \times i$ minors of M and vice versa (exercise). Hence

$$\Delta_i^M | \Delta_i^S \quad \text{and} \quad \Delta_i^S | \Delta_i^M,$$

s.t.

$$\Delta_i^M = \Delta_i^S =: \Delta_i$$

or equiv. the monic g.c.d. of all $i \times i$ minors is invariant under equivalence. Now from the form of S in (19) and division property (21) $\Delta_i = \prod_{j=1}^{i} \lambda_j$ for $i \in \underline{r}$; hence, with $\Delta_0 \equiv 1$, (23) is established. Since the Δ_i are unique, by (23) the λ_i must also be unique and we are done. ∎

29 Exercise. Find the Smith form of

$$M(s) = \begin{bmatrix} s+1 & 0 & s(s+1) \\ 0 & s+2 & s(s+2) \end{bmatrix}$$

30 Exercise. Show that the Smith form of a unimodular matrix is the unit matrix. (This is used in the proof of Theorem 2.3.2.13.)

From the proof of Theorem 19 and from Exercise 2.3.2.16, we have also

31 Corollary [Characterization of equivalence]. Two polynomial matrices M and $N \in \mathbb{R}[s]^{m \times n}$ are equivalent, equiv. can be obtained from each other by e.o.'s, if and only if one the following conditions hold:

(a) M and N have the same Smith form;

(b) M and N have the same invariant polynomials;

(c) M and N have the same determinantal divisors. ∎

2.3.5. The Solution Space of $D(p)\xi(t) = \theta_\nu$, $t \geq 0$

In this section we continue the description of the $\mathbb{R}$-linear space X of solutions $\xi(\cdot) : (0-, \infty) \to \mathbb{R}^\nu$ of the differential equation

1 $$D(p)\xi(t) = \theta_\nu \quad t \geq 0,$$

where $D(\cdot) \in \mathbb{R}[p]^{\nu \times \nu}$ is nonsingular and $p = d/dt$ is the differential operator: see Sec. 2.3.3.

A first result involves the dimension of the solution space X.

2 Theorem [Dimension of the solution space]. Let $D(\cdot) \in \mathbb{R}[p]^{\nu \times \nu}$ be nonsingular and consider the $\mathbb{R}$-linear space X of solutions $\xi(\cdot) : (0-, \infty) \to \mathbb{R}^\nu$ of the differential equation

1 $$D(p)\xi(t) = \theta_\nu \quad t \geq 0$$

U.t.c.

3 $$\dim X = \partial[\det D].$$ ∎

4 Comment. (3) means that any solution of (1) is an $\mathbb{R}$-linear combination of a basis consisting of $n := \partial[\det D]$ linearly independent solutions: once a basis of X is known then all solutions are known.

5 Proof. From Theorem 2.3.3.26 and Comment 2.3.3.31, without loss of generality, we may assume that $D(\cdot)$ is in the Smith form of Theorem 2.3.4.18, i.e.,

6 $$D(p) = S(p) = \mathrm{diag}[\lambda_i(p)]_{i=1}^{\nu},$$

where the $\lambda_i(\cdot) \in \mathbb{R}[p]$ are the invariant polynomials of $D(\cdot)$. As a consequence with $\xi(\cdot) = (\xi_i(\cdot))_{i=1}^{\nu}$, $\xi_i(\cdot) : (0-, \infty) \to \mathbb{R}$, $\forall i \in \underline{\nu}$, the vector differential equation (1) decouples into ν scalar differential equations $\lambda_i(p)\xi_i(t) = 0$, $t \geq 0$ and the solution space X is the direct sum of component spaces X_i, $i \in \underline{\nu}$. More precisely,

7 $$X = \bigoplus_{i=1}^{\nu} X_i,$$

where

$$X_i = \{\xi(\cdot) : (0-, \infty) \to \mathbb{R}^{\nu};$$

$$\xi(\cdot) = (0, \cdots, 0, \underset{\substack{\uparrow \\ i\text{th component}}}{\xi_i(\cdot)}, 0, \cdots, 0)^T,$$

$$\lambda_i(p)\xi_i(t) = 0 \quad \forall t \geq 0\}$$

Now by the theory of scalar differential equations

8 $$\dim X_i = \partial[\lambda_i(\cdot)].$$

(A basis for X_i is generated by $\partial[\lambda_i(\cdot)]$ linearly independent solutions of $\lambda_i(p)\xi_i(t) = 0$, $t \geq 0$.) Hence by (6)-(8)

$$\dim X = \sum_{i=1}^{\nu} \dim X_i = \sum_{i=1}^{\nu} \partial[\lambda_i] = \partial[\det D].$$ ■

From the structure of vector spaces and Theorem 2 we have immediately

11 Theorem. Let $D(\cdot) \in \mathbb{R}[p]^{\nu\times\nu}$ be nonsingular and consider the differential equation

1 $$D(p)\xi(t) = \theta_\nu \quad t \geq 0$$

with solution space X.

Let

12 $$n := \partial[\det D]$$

denote the dimension of X and let $(\psi_k(\cdot))_{k=1}^{n}$ denote any basis of X generating a basis matrix $\Psi(\cdot) : (0-, \infty) \to \mathbb{R}^{\nu\times n}$ given by

13 $$\Psi(t) = \left[\, \psi_1(t) \,\vdots\, \psi_2(t) \,\vdots\, \cdots \,\vdots\, \psi_n(t) \,\right], \text{ for } t > 0-.$$

Let $x(0) = (x_k(0))_{k=1}^{n}$ be any n-vector of $\mathbb{R}^n$

U.t.c.

(a) $\forall\, \xi(\cdot) \in X$, or equiv. $\forall$ solution of (1), there is a unique $x(0) \in \mathbb{R}^n$ s.t.

14 $$\xi(t) = \Psi(t)x(0), \quad \forall t > 0-,$$

(b) the map

15 $$x(0) \in \mathbb{R}^n \mapsto \xi(\cdot) = \Psi(\cdot)x(0) \in X$$

is an $\mathbb{R}$-linear bijection (isomorphism) from $\mathbb{R}^n$ onto X. ■

Theorem 11 justifies the following definition.

16 Definition. We call state at $t = 0$ of differential equation (1) the n-vector $x(0) \in \mathbb{R}^n$ defined by equation (14) through the choice of a basis $(\psi_k(\cdot))_{k=1}^{n}$ for the solution space X.

17 Comments. (a) For every choice of basis the definition of the state at $t = 0$ is justified because of map (15): the knowledge of $x(0)$ completely determines the trajectory $\xi(\cdot)$; conversely to every trajectory corresponds a unique state $x(0)$ at $t = 0$.

(b) The definition of the state at $t = 0$ depends on the choice of a basis for the solution space $\mathcal{X}$ of (1): below we shall normalize this definition by finding a state at $t = 0$ related to initial data of $\xi(\cdot)$ and its derivatives: see Algorithm 36.

(c) For every choice of basis the dimension of the state space is

12 $$n := \partial[\det D].$$

(d) $x(0)$ is zero means that $\xi(\cdot)$ and all its derivatives are identically zero.
(e) For two arbitrary bases $(\psi_k(\cdot))_{k=1}^{n}$ and $(\bar{\psi}_k(\cdot))_{k=1}^{n}$ of the solution space $\mathcal{X}$ of (1), the relation between the basis matrices $\Psi(\cdot)$ and $\bar{\Psi}(\cdot)$ (defined as in (13)) is given by

18 $$\Psi(\cdot) = \bar{\Psi}(\cdot)P,$$

where $P \in \mathbb{R}^{n\times n}$ is nonsingular. As a consequence the states $x(0) \in \mathbb{R}^n$ and $\bar{x}(0) \in \mathbb{R}^n$ defined by

19 $$\xi(t) = \Psi(t)x(0) = \bar{\Psi}(t)\bar{x}(0), \quad \text{for } t > 0-,$$

are related by

20 $$\bar{x}(0) = P\,x(0),$$

where $P \in \mathbb{R}^{n\times n}$ is nonsingular.

(f) A convenient method to obtain a basis $(\psi_k(\cdot))_{k=1}^{n}$ of the solution space $\mathcal{X}$ of (1) is to reduce first $D(p)$ to the upper-triangular Hermite row form by Algorithm 2.3.4.5.L (leaving each solution $\xi(\cdot)$ and the vector space $\mathcal{X}$ unchanged; see Theorem 2.3.3.6), and then use Algorithm 24 below.

24 Algorithm [Construction of a basis of solutions]
Data: $D = [d_{ij}]_{i,j\in\underline{\nu}} \in \mathbb{R}[p]^{\nu\times\nu}$ is nonsingular and in upper-triangular Hermite row form as described by Theorem 2.3.4.1.L and Corollary 2.3.4.9.L.

We consider a differential equation

1 $$D(p)\xi(t) = \theta_\nu \quad t \geq 0,$$

or equiv. a system of differential equations

25 $$\sum_{j=i}^{\nu} d_{ij}(p)\xi_j(t) = 0 \quad t \geq 0 \quad \text{for } i \in \underline{\nu}$$

with solution space X. Define

26 $$n_i := \partial[d_{ii}(\cdot)] \quad \text{for } i \in \underline{\nu}.$$

<u>Objective</u>: Construct a basis $(\psi_k(\cdot))_{k=1}^{n}$ of X, $n := \partial[\det D]$.

<u>Procedure</u>: For $i = 1 \sim \nu$ s.t. $n_i \neq 0$, do:

<u>Step 1</u>. Find n_i linearly independent solutions $\xi_i(\cdot) : [0, \infty) \to \mathbb{R}$ of the <u>scalar</u> differential equation

27 $$d_{ii}(p)\xi_i(t) = 0, \quad t \geq 0.$$

<u>Step 2</u>. If $i = 1$, skip step 2.
Else, for every linearly independent solution $\xi_i(\cdot)$ of step 1, for every $j = i-1, i-2, \cdots, 1$, solve for $\xi_j(\cdot)$ the scalar differential equation

28 $$d_{jj}(p)\xi_j(t) + d_{j(j+1)}\xi_{j+1}(t) + \cdots + d_{ji}(p)\xi_i(t) = 0$$

with initial conditions zero and driven by $\xi_i(\cdot), \xi_{i-1}(\cdot), \cdots, \xi_{j+1}(\cdot)$.
<u>Step 3</u>. From every independent solution $\xi_i(\cdot)$ of step 1 together with its corresponding set of solutions $\xi_j(\cdot)$ of step 2, (empty if $i = 1$), form a new basis vector of X given by

29 $$\psi_k(t) = \begin{bmatrix} \xi_1(t) \\ \vdots \\ \xi_{i-1}(t) \\ \xi_i(t) \\ \vdots \\ 0 \end{bmatrix} \in \mathbb{R}^\nu, \; t > 0-.$$

End of Algo.

30 Example. Consider the following example where n = 3.

$$D(p)\xi(t) = \begin{bmatrix} p+1 & p+1 \\ 0 & (p+2)^2 \end{bmatrix} \begin{bmatrix} \xi_1(t) \\ \xi_2(t) \end{bmatrix} = \theta_2 \qquad t \geq 0.$$

$i = 1$: $n_1 = 1$: $\xi_1(t) = e^{-t}$,

$$\psi_1(t) = \begin{bmatrix} e^{-t} \\ 0 \end{bmatrix};$$

$i = 2$: $n_2 = 2$: (a) $\xi_2(t) = e^{-2t}$, $\xi_1(t) = e^{-t} - e^{-2t}$,

$$\psi_2(t) = \begin{bmatrix} e^{-t} - e^{-2t} \\ e^{-2t} \end{bmatrix};$$

(b) $\xi_2(t) = te^{-2t}$, $\xi_1(t) = -te^{-2t}$,

$$\psi_3(t) = \begin{bmatrix} -te^{-2t} \\ te^{-2t} \end{bmatrix}.$$

STOP. A basis for X is $(\psi_k(\cdot))_{k=1}^{3}$.

31 Remark. $n_i := \partial[d_{ii}(\cdot)] = 0$ means that no contribution is made to the solution space X: the solution of the scalar differential equation $d_{ii}(p)\xi_i(t) = 0$ $t \geq 0$ is $\xi_i(t) = 0$ for $t > 0-$; moreover, $\forall j = i-1, i-2, \cdots, 1$, the "driven" differential equation (28) is not driven: hence with initial conditions zero, $\forall j < i, \xi_j(t) = 0$ $\forall t > 0-$ $\bullet\bullet\bullet$.

32 Exercise. Consider the differential equation (1). Show that the reduction of $D(\cdot)$ to Hermite row form and the subsequent application of Algorithm 24 leads to the construction of a basis for the solution space X.

33 Remark. When applying the Laplace transform method in Algorithm 24 be sure to take into account the initial values of functions and their derivatives when needed.

34 Exercise. Compute a basis for the solution space X of the differential equation (1) where $D(p)$ is given by

$$D(p) = \begin{bmatrix} 1 & p & p^2 \\ 0 & p+1 & p \\ 0 & 0 & p^2 \end{bmatrix},$$

$$D(p) = \begin{bmatrix} 1 & p+1 & p^2 \\ 0 & 1 & p \\ 0 & 0 & 1 \end{bmatrix},$$

and

$$D(p) = \begin{bmatrix} (p+1)^2 & p \\ 1 & p+2 \end{bmatrix}.$$

An immediate consequences to Algorithm 24 is:

36 Algorithm [Construction of a state at $t = 0$ using initial data]
Data: we are given the differential equation (1) where $D(\cdot) \in \mathbb{R}[p]^{\nu \times \nu}$ is nonsingular and $n := \partial[\det D]$.
Step 1. By Algorithm 2.3.4.5.L reduce $D(\cdot)$ to the upper-triangular Hermite row form.
Step 2. For every $i \in \underline{\nu}$ compute n_i, the degree of entry (i, i) of the Hermite form.
Step 3. For every $i \in \underline{\nu}$ s.t. $n_i \neq 0$ compute the initial data

$$\xi_i(0),\ \xi_i^{(1)}(0),\ \cdots,\ \xi_i^{(n_i-1)}(0)$$

(denoting derivates of $\xi_i(\cdot)$ at 0), and stack them into an n-vector $x_\xi(0) \in \mathbb{R}^n$.

End of Algo.

37 Example. Consider the differential equation of Example 30: $n_1 = 1$, $n_2 = 2$; hence

$$x_\xi(0) = \begin{bmatrix} \xi_1(0) \\ \xi_2(0) \\ \xi_2^{(1)}(0) \end{bmatrix}.$$

(As an exercise show that $t \mapsto x_\xi(t)$ satisfies $\dot{x}_\xi = Ax_\xi$ for some $A \in \mathbb{R}^{3\times 3}$).

38 Theorem. The n-vector $x_\xi(0) \in \mathbb{R}^n$ constructed in Algorithm 36 is a state at $t = 0$ for the differential equation (1), or equiv. there exists a basis $(\bar{\psi}_k(\cdot))_{k=1}^n$ for the solution space X of (1) such that with the basis matrix $\bar{\Psi}(\cdot)$ (defined as in (13)),
(a) $\forall\ \xi(\cdot) \in X$, or equiv. $\forall$ solution of (1), there is a unique $x_\xi(0) \in \mathbb{R}^n$ s.t.

38a $$\xi(t) = \bar{\Psi}(t)x_\xi(0) \quad \text{for } t > 0-$$

(b) the map

38b $$x_\xi(0) \in \mathbb{R}^n \mapsto \xi(\cdot) = \bar{\Psi}(\cdot)x_\xi(0) \in X$$

is an $\mathbb{R}$-linear bijection (isomorphism) of $\mathbb{R}^n$ onto X. Moreover, let $(\psi_k(\cdot))_{k=1}^n$ be any basis of the solution space X of (1), with basis matrix $\psi(\cdot)$ defined by (13), and consider for differential equation (1) $x(0) \in \mathbb{R}^n$, viz., the state at $t = 0$ defined by equation (14).
U.t.c.
there exists a nonsingular matrix $P \in \mathbb{R}^{n\times n}$ s.t. the map

39 $$x_\xi(0) \in \mathbb{R}^n \mapsto x(0) = Px_\xi(0) \in \mathbb{R}^n$$

is a linear bijection. As a consequence every solution $\xi(\cdot)$ of (1) has the form

40 $$\xi(t) = \psi(t)Px_\xi(0) \quad \text{for } t > 0-. \qquad ■$$

41 Exercise. Prove Theorem 38. [Hints: (a) Use Algorithm 24 and the fact that the solution $\xi_i(\cdot)$ of any nontrivial scalar differential equation (27) is uniquely determined by $\xi_i(0)$, $\xi_i^{(1)}(0)$, $\cdots$, $\xi^{(n_i-1)}(0)$ with $n_i = \partial[d_{ii}(\cdot)]$; (b) use (19)-(20).]

42 Comment. The state $x_\xi(0) \in \mathbb{R}^n$ constructed by Algorithm 36 is specified by $\xi(0)$ and its derivatives evaluated at 0. For a precise formulation, see step 3 of Algorithm 36. Note in particular that if $n_i = 0$, $\xi_i(0)$ does not appear in $x_\xi(0)$.

At this point it is convenient to introduce the following definitions.

43 Definitions. Consider the differential equation (1). We call normalized state at $t = 0$ of (1) or simply the state at $t = 0$ of (1) the n-vector $x_\xi(0)$ constructed by Algorithm 36. We call (normalized) state trajectory of (1) the map $t \mapsto x_\xi(t)$. Finally, we call pseudo-state at $t = 0$ and pseudo-state trajectory of (1) $\xi(0)$ resp. the map $t \mapsto \xi(t)$.

44 Comments. (a) $x_\xi(0) = \theta_n$ is equivalent to $\xi(\cdot)$ and all its derivatives are zero.
(b) In general, $\xi(0) \neq x_\xi(0)$ and $\nu = \dim \xi(0) \neq n = \dim x_\xi(0) = \partial[\det D]$: we may have $\nu < n$ or $\nu = n$ or $\nu > n$.
(c) $\xi(t)$ is called pseudo-state because $\xi(0)$ and some of its derivates at 0 determine the state $x_\xi(0)$. In some cases the map $x_\xi(0) \mapsto \xi(0)$ is not surjective,(see (40) with $t = 0$).

45 Exercise. Consider differential equation (1) with $D(p) = pI - A$, where $A \in \mathbb{R}^{\nu\times\nu}$. (a) Show that $\nu = n = \partial[\det D]$ but in general $\xi(0) \neq x_\xi(0)$, (e.g., $A = \begin{bmatrix} 1 & 3 \\ 2 & 2 \end{bmatrix}$ leads to $x_\xi(0) = (\xi_2(0), \xi_2^{(1)}(0))^T$. (b) Knowing that $\xi(t) = \exp(At)\xi(0)$ for $t > 0-$, show that there exists a basis for X s.t. $x(0) = \xi(0)$, hence $\xi(0)$ is a state at $t = 0$.

We conclude this section with two important results. The first result characterizes trivial differential equations. The second result displays the invariance of the zero-state response of a PMD under e.o.'s.

50 Fact [Trivial dynamics]. Let $D(\cdot) \in \mathbb{R}[p]^{\nu\times\nu}$ be nonsingular and consider the differential equation

1 $$D(p)\xi(t) = \theta_\nu \quad t \geq 0$$

with solution space X.
U.t.c.

51 $$X = \{\theta\}$$

where θ stands for the solution $\xi(t) = \theta_\nu\ \forall t > 0-$,

⇔

$D(\cdot)$ is unimodular.

52 Exercise. Prove Fact 50.

54 Theorem [Invariance of the z-s response of a PMD under e.o.'s].
Consider the PMD $[D, N_\ell, N_r, K]$ described by

55
$$\begin{aligned} D(p)\xi(t) &= N_\ell(p)u(t) \\ y(t) &= N_r(p)\xi(t) + K(p)u(t) \end{aligned} \qquad t \geq 0$$

where (a) $D(\cdot)$, $N_\ell(\cdot)$, $N_r(\cdot)$, $K(\cdot) \in E(\mathbb{R}[p])$ and have dimensions $\nu \times \nu$, $\nu \times n_i$, $n_o \times \nu$, $n_o \times n_i$; moreover $D(\cdot)$ is nonsingular.
(b) $u(\cdot) : \mathbb{R}_+ \to \mathbb{R}^{n_i}$, $\xi(\cdot) : \mathbb{R}_+ \to \mathbb{R}^\nu$, $y(\cdot) : \mathbb{R}_+ \to \mathbb{R}^\nu$ are the input, pseudo-state, and output of the PMD.
(c) the z-s response $y(\cdot)$ is obtained by setting to zero the n-vector $x_\xi(0)$ (constructed in Algorithm 36), and driving the PMD by the input $u : \mathbb{R}_+ \to \mathbb{R}^{n_i}$, where by assumption $u(\cdot)$ and all its derivatives have zero value at $t = 0-$.

Let now

56
$$\begin{aligned} \bar{D}(p) &:= L(p)D(p)R(p) & \bar{N}_\ell(p) &:= L(p)N_\ell(p) \\ \bar{N}_r(p) &:= N_r(p)R(p) & \bar{K}(p) &:= K(p) \end{aligned}$$

where $L(\cdot)$ and $R(\cdot)$ are unimodular matrices representing e.o.'s over $\mathbb{R}[p]$.
Consider then the transformed PMD $[\bar{D}, \bar{N}_\ell, \bar{N}_r, \bar{K}]$ described by

57
$$\left.\begin{aligned} \bar{D}(p)\bar{\xi}(t) &= \bar{N}_\ell(p)u(t), \\ y(t) &= \bar{N}_r(p)\bar{\xi}(t) + \bar{K}(p)u(t) \end{aligned}\right\} t \geq 0 .$$

U.t.c.

$\forall\, u(\cdot) : \mathbb{R}_+ \to \mathbb{R}^{n_i}$ piecewise sufficiently differentiable, the z-s responses of the PMD's (55) and (57) satisfy

$$y(t) = \bar{y}(t) \qquad \forall t > 0-.$$ ■

58 Exercise. Prove Fact 54. (Hint: Use the Laplace transform method where, at 0-, the values of $\xi(\cdot)$, $\bar{\xi}(\cdot)$, $u(\cdot)$, and their derivatives are zero by assumption.)

2.3.6. Greatest Common Divisor Extraction

We handle the case of a g.c.r.d.

1.R Algorithm [Greatest common right divisor extraction]

Data: $\bar{N}_r \in \mathbb{R}[s]^{n_o \times n_i}$, $\bar{D}_r \in \mathbb{R}[s]^{n_i \times n_i}$ with $\bar{D}_r$ nonsingular.

Step 1. Set

2 $$M := \begin{matrix} n_i \\ n_o \end{matrix}\overset{n_i}{\begin{bmatrix} \bar{D}_r \\ \hline \bar{N}_r \end{bmatrix}} \in E(\mathbb{R}[s])$$

and observe that M has full column rank over $\mathbb{R}[s]$.

Step 2. Use Algorithm 2.3.4.5.L to get M in upper triangular form by e.r.o.'s: see Corollary 2.3.4.9.L: step 6 of the algorithm may be skipped.

As a consequence we obtain unimodular matrices W and $W^{-1} \in E(\mathbb{R}[s])$ with

3 $$W := \underset{n_i \quad n_o}{\begin{bmatrix} V_r & U_r \\ -N_\ell & D_\ell \end{bmatrix}}\begin{matrix} n_i \\ n_o \end{matrix} \qquad W^{-1} := \underset{n_i \quad n_o}{\begin{bmatrix} D_r & -U_\ell \\ N_r & V_\ell \end{bmatrix}}\begin{matrix} n_i \\ n_o \end{matrix}$$

such that

4 $$WM = \overset{n_i \quad n_o}{\begin{bmatrix} V_r & U_r \\ -N_\ell & D_\ell \end{bmatrix}} \overset{n_i}{\begin{bmatrix} \bar{D}_r \\ \bar{N}_r \end{bmatrix}}\begin{matrix} n_i \\ n_o \end{matrix} = \begin{bmatrix} R \\ 0 \end{bmatrix}\begin{matrix} n_i \\ n_o \end{matrix},$$

where

R is upper-triangular and nonsingular. ■

5 Remarks. (a) The unimodular matrices W and W^{-1} in (3) are obtained simultaneously by starting with a unit matrix $I_{n_o+n_i}$ for both W and W^{-1} and

performing upon it the e.r.o.'s of step 2 for getting W and their inverses as e.c.o.'s for getting W^{-1}: no inversion of W is needed.
The translation rules for e.r.o.'s and their inverses as e.c.o.'s are

1. $\rho_i \rightleftarrows \rho_j \quad : \gamma_i \rightleftarrows \gamma_j$,

2. $\rho_i \leftarrow k\rho_i \quad : \gamma_i \leftarrow k^{-1}\gamma_i$,

3. $\rho_i \leftarrow \rho_i + r\rho_j : \gamma_j \leftarrow \gamma_j - r\gamma_i$.

(b) The reason for the special notation of the partitions of W and W^{-1} in (3) will become clear below.

6 Exercise. Apply Algorithm 1.R to

$$\bar{N}_r(s) = [(s+1)^2 \quad s-1] \,,\; \bar{D}_r(s) = \begin{bmatrix} s+1 & s-1 \\ 0 & (s-1)^2 \end{bmatrix},$$

and show that a result is

$$W = \left[\begin{array}{cc|c} -s & 1 & 1 \\ s+1 & -1 & -1 \\ \hline 1-s^2 & s & s-1 \end{array}\right], \quad W^{-1} = \left[\begin{array}{cc|c} 1 & 1 & 0 \\ 0 & s-1 & 1 \\ \hline s+1 & 1 & -1 \end{array}\right]$$

$$\begin{bmatrix} R \\ \hline 0 \end{bmatrix} = \left[\begin{array}{cc} s+1 & 0 \\ 0 & s-1 \\ \hline 0 & 0 \end{array}\right].$$

10.R Fact. Let $(\bar{N}_r, \bar{D}_r) \in \mathbb{R}[s]^{n_o \times n_i} \times \mathbb{R}[s]^{n_i \times n_i}$ with $\bar{D}_r$ nonsingular and apply Algorithm 1.R. Then the upper triangular matrix R in (4) is a g.c.r.d. of $(\bar{N}_r, \bar{D}_r)$.

Proof. (a) By (4)

$$W^{-1}\begin{bmatrix} R \\ \hline 0 \end{bmatrix} = \begin{bmatrix} \bar{D}_r \\ \hline \bar{N}_r \end{bmatrix}.$$

Hence by (3^2) $\bar{D}_r = D_r R$, $\bar{N}_r = N_r R$. Hence R is a c.r.d. of $(\bar{N}_r, \bar{D}_r)$.

(b) By (4) again, $U_r\bar{N}_r + V_r\bar{D}_r = R$. Hence any c.r.d. of $(\bar{N}_r, \bar{D}_r)$ is a r.d. of R, or equiv. R is a ℓ.m. of every c.r.d. of $(\bar{N}_r, \bar{D}_r)$. ∎

Recall now the definition of left-equivalence: see Sec. 2.3.2.

11.R <u>Fact</u>. Let $(\bar{N}_r, \bar{D}_r) \in \mathbb{R}[s]^{n_o \times n_i} \times \mathbb{R}[s]^{n_i \times n_i}$ with $\bar{D}_r$ nonsingular. Then all the g.c.r.d.'s of $(\bar{N}_r, \bar{D}_r)$ are <u>left</u>-equivalent, or equiv. if R' and R" are g.c.r.d.'s of $(\bar{N}_r, \bar{D}_r)$; then

$$R' \overset{\ell}{\sim} R''.$$

12 <u>Comment</u>. All g.c.r.d.'s are related by e.r.o.'s or equiv. or a <u>g.c.r.d.</u> is <u>unique up to a left unimodular factor</u>.

<u>Proof</u>. Because left equivalence is an equivalence relation and by Fact 10.R matrix R of (4) is a g.c.r.d. of $(\bar{N}_r, \bar{D}_r)$, it is sufficient to show that

13 $$R' \underset{\sim}{\ell} R$$

for any R' a g.c.r.d. of $(\bar{N}_r, \bar{D}_r)$. Now since R and R' are g.c.r.d.'s of $(\bar{N}_r, \bar{D}_r)$, they are c.r.d.'s of $(\bar{N}_r, \bar{D}_r)$ and R must be a ℓ.m. of R' and vice versa, i.e., there exist square matrices L and $L' \in E(\mathbb{R}[s])$ s.t.

14 $$R' = L'R \qquad R = LR'.$$

Hence $R = LL'R$ with R nonsingular by (4). Hence $I = LL'$ such that L and L' are unimodular with $L' = L^{-1}$. Hence (13) holds. ∎

15.R <u>Fact</u>. Let $(N_r, D_r) \in \mathbb{R}[s]^{n_o \times n_i} \times \mathbb{R}[s]^{n_i \times n_i}$ with D_r nonsingular. Then

(N_r, D_r) is r.c.

⇔

all g.c.r.d.'s of (N_r, D_r) are unimodular.

<u>Proof</u>. Because of Fact 11.R if one g.c.r.d. is unimodular, then all g.c.r.d.'s are unimodular. ∎

We give now two characterizations of right-coprimeness.

16.R <u>Theorem</u> [Bezout identity]. Let $(N_r, D_r) \in \mathbb{R}[s]^{n_o \times n_i} \times \mathbb{R}[s]^{n_i \times n_i}$ with D_r nonsingular. Then

(N_r, D_r) is r.c.

$\Leftrightarrow$

$\exists\; U_r, V_r \in E(\mathbb{R}[s])$ s.t.

17 $$U_r N_r + V_r D_r = I_{n_i}.$$

18 <u>Comment</u>. Equation (17) is called the <u>Bezout identity</u> [Kai.1,p.379]. It generalizes to the multivariable case the well-known Bezout identity un + vd = 1 for coprime polynomials (n, d) [Sig.1]. Notice also that condition (17) shows that $[V_r\; U_r]$ is a left inverse of the matrix $\begin{bmatrix} D_r \\ -- \\ N_r \end{bmatrix}$ where all operations are within the ring $\mathbb{R}[s]$.

19 <u>Proof of Theorem 16.R</u>. (a) $\Rightarrow$: Use Algorithm 1.R with $\bar{N}_r \leftarrow N_r$ and $\bar{D}_r \leftarrow D_r$. Then one obtains by (4) $U_r\bar{N}_r + V_r\bar{D}_r = R$ with R a g.c.r.d. of $(\bar{N}_r, \bar{D}_r)$ by Fact 10.R. Now $(\bar{N}_r, \bar{D}_r)$ is a right coprime pair. Hence, by Fact 15.R, R must be unimodular. Therefore, by premultiplication by R^{-1} one obtains $\bar{U}_r\bar{V}_r + \bar{V}_r\bar{D}_r = I_{n_i}$, where $\bar{U}_r = R^{-1}U_r$ and $\bar{V}_r = R^{-1}V_r \in E(\mathbb{R}[s])$: we have a Bezout identity.

(b) $\Leftarrow$: Because of (17) any g.c.r.d. of (N_r, D_r) is unimodular. ■

20.R <u>Theorem</u> [Rank test]. Let $(N_r, D_r) \in \mathbb{R}[s]^{n_o \times n_i} \times \mathbb{R}[s]^{n_i \times n_i}$ with D_r nonsingular.

U.t.c.

(N_r, D_r) is r.c.

$\Leftrightarrow$

21 $$\operatorname{rk} \overset{n_i}{\begin{bmatrix} D_r(s) \\ ----- \\ N_r(s) \end{bmatrix}} \begin{matrix} n_i \\ \\ n_o \end{matrix} = n_i \quad \forall s \in \mathbb{C}.$$

22 <u>Comment</u>. Condition (21) is known as the <u>rank test</u>.

(a) It generalizes to the multivariable case the well-known condition that a polynomial pair (n, d) is coprime iff they have no common zeros, i.e., iff $\mathrm{rk}[d(s)\ n(s)]^T = 1$, $\forall s \in \mathbb{C}$, i.e., iff they have no noninvertible common factors. For our matrix case this reads: (N_r, D_r) is r.c. iff they have no nonunimodular right common factors.

(b) From Comment (18) it follows also: the full column rank matrix $\begin{bmatrix} D_r \\ \hline N_r \end{bmatrix}$ is left invertible over $\mathbb{R}[s]$ iff the rank test (21) holds.

23 Exercise. Use the rank test (21) to show that $(\bar{N}_r, \bar{D}_r)$ as given in Exercise 6 is not r.c. Find a nonunimodular right common factor.

24 Proof of Theorem 20.R. Apply Algorithm 1.R with $\bar{N}_r \leftarrow N_r$ and $\bar{D}_r \leftarrow D_r$. According to (4) and Fact 10.R one obtains

$$W\begin{bmatrix} \bar{D}_r \\ \hline \bar{N}_r \end{bmatrix} = \begin{bmatrix} R \\ \hline 0 \end{bmatrix},$$

where (a) W is unimodular and represents e.r.o.'s; (b) R is a g.c.r.d. of $(\bar{N}_r, \bar{D}_r)$. Observe now that e.r.o.'s result in rank invariance at every $s \in \mathbb{C}$ and that by Fact 15.R (N_r, D_r) is r.c. iff R is unimodular. Hence

$$\mathrm{rk}\begin{bmatrix} D_r(s) \\ N_r(s) \end{bmatrix} = n_i,\ \forall s \in \mathbb{C} \iff R \text{ is unimodular} \iff (N_r, D_r) \text{ is r.c.}$$ ∎

27 Convention. Algorithm 1.R extracts a g.c.r.d. $R \in \mathbb{R}[s]^{n_i \times n_i}$ from $(\bar{N}_r, \bar{D}_r) \in \mathbb{R}[s]^{n_o \times n_i} \times \mathbb{R}[s]^{n_i \times n_i}$ with $\bar{D}_r$ nonsingular. Consider now $(\bar{D}_\ell, \bar{N}_\ell) \in \mathbb{R}[s]^{n_o \times n_o} \times \mathbb{R}[s]^{n_o \times n_i}$ with $\bar{D}$ nonsingular. Then similarly by e.c.o.'s on

28 $$M = \overset{\quad n_o \quad n_i}{[\bar{D}_\ell \,\vdots\, \bar{N}_\ell]}\, n_o$$

one can get M in lower triangular form by Algorithm 2.3.4.5.R and Corollary 2.3.4.9.R: see Convention 2.3.4.12. As a consequence there exists a unimodular matrix $W \in E(\mathbb{R}[s])$ s.t.

29 $$MW =: [\bar{D}_\ell \mid \bar{N}_\ell] \overset{\quad n_o \quad n_i}{\begin{bmatrix} V_\ell & -N_r \\ \hline U_\ell & D_r \end{bmatrix}} = [L \mid 0]\, n_o ,$$

where

L is <u>lower-triangular</u> and <u>nonsingular</u>.

End of Procedure

The procedure above is called Algorithm 1.L. [Extraction of a g.c.ℓ.d.]. Its consequences are similar to those of Algorithm 1.R, viz., Fact 10.L (L in (29) is a g.c.ℓ.d. of $(\bar{D}_\ell, \bar{N}_\ell)$). Fact 11.L (all g.c.$\ell$.d.'s of $(\bar{D}_\ell, \bar{N}_\ell)$ are right-equivalent), Fact 15.L ($(\bar{D}_\ell, \bar{N}_\ell)$ is ℓ.c. iff every g.c.ℓ.d. is unimodular), Theorem 16.L (Bezout identity for left coprimeness), Theorem 20.L (rank test for left coprimeness). Notice that we have used R (resp. L) to indicate a result on right extraction, right coprimeness$\cdots$(resp. left extraction, left coprimeness): we adopt the following convention:

Throughout the text a suffix R (resp. L) will mean that we are referring to a procedure or result which has a <u>dual analog</u> indicated by a suffix L (resp. R) which should be recognizable from the context.

30 <u>Exercise</u>. Draw a comparative table with in the first column Facts 10.R, 11.R, 15.R, Theorems 16.R, 20.R and in the second column Facts 10.L, 11.L, 15.L, Theorems 16.L, 20.L.

2.4. Matrix Fraction Descriptions of Rational Transfer Function Matrices

In this section we consider rational transfer functions as polynomial matrix fractions.

2.4.1. Coprime Fractions

1.R <u>Definitions</u>. Let $H \in \mathbb{R}(s)^{n_o \times n_i}$. We say that $(N_r, D_r) \in \mathbb{R}[s]^{n_o \times n_i} \times \mathbb{R}[s]^{n_i \times n_i}$ is a <u>right coprime fraction</u> (r.c.f.) of H iff

(a) $\det D_r \not\equiv 0$,

(b) $H = N_r D_r^{-1}$,

(c) (N_r, D_r) is r.c.

If (c) is not required we say that (N_r, D_r) is a <u>right fraction</u> (r.f.) of H. ∎

1.L <u>Definitions</u>. Let $H \in \mathbb{R}(s)^{n_o \times n_i}$. We say that $(D_\ell, N_\ell) \in \mathbb{R}[s]^{n_o \times n_o} \times \mathbb{R}[s]^{n_o \times n_i}$ is a <u>left coprime fraction</u> (ℓ.c.f.) of H iff

(a) $\det D_\ell \not\equiv 0$,

(b) $H = D_\ell^{-1} N_\ell$,

(c) (D_ℓ, N_ℓ) is ℓ.c.

If (c) is not required, we say that (D_ℓ, N_ℓ) is a <u>left fraction</u> (ℓ.f.) of H. ∎

In the following we consider in detail right fractions; we leave it to the reader to investigate left fractions.

2.R <u>Algorithm</u> [Search of a r.c.f.]

<u>Data</u>: $H \in \mathbb{R}(s)^{n_o \times n_i}$.

<u>Step 1</u>. $\forall j \in \underline{n}_i$ compute

3 d_j := a least common denominator (ℓ.c.d.) of all entries of γ_j, where

$d_j := 1$ if γ_j is zero.

<u>Step 2</u>. $\forall i \in \underline{n}_o$ and $\forall j \in \underline{n}_i$ write every entry h_{ij} of H as

4 $$h_{ij} = \bar{n}_{ij}/d_j .$$

Set

5 $$\bar{N}_r := [\bar{n}_{ij}], \quad \bar{D}_r := \mathrm{diag}[d_j]_{j=1}^{n_i} .$$

<u>Comment</u>: $(\bar{N}_r, \bar{D}_r)$ is a r.f. of H.

<u>Step 3</u>. Using Algorithm 2.3.6.1.R, extract a g.c.r.d. $R \in \mathbb{R}[s]^{n_i \times n_i}$ of $(\bar{N}_r, \bar{D}_r)$. As a result:

6 $\qquad \bar{N}_r = N_r R \quad \bar{D}_r = D_r R.$

Comment: (N_r, D_r) is a r.c.f. of H.

End of Algo

7 Remarks. (a) Steps 1 and 2 may be replaced by any method which delivers a r.f. $(\bar{N}_r, \bar{D}_r)$ of H.
(b) The application of extraction Algorithm 2.3.6.1.R in step 3 delivers matrices N_r, D_r of a r.c.f. as submatrices of the partition of unimodular matrix W^{-1} in (2.3.6.3). See the proof of Fact 8.R.

8.R Fact. Every $H \in \mathbb{R}(s)^{n_o \times n_i}$ has a r.c.f. (N_r, D_r).

Proof. Use Algorithm 2.R and observe that in step 3 (6) holds since (2.3.6.3) and (2.3.6.4) result in

$$W^{-1}\begin{bmatrix} R \\ \hline 0 \end{bmatrix} = \begin{bmatrix} \bar{D}_r \\ \hline \bar{N}_r \end{bmatrix}.$$

Moreover, since $WW^{-1} = I$, a Bezout identity is obtained from the partitions of W and W^{-1} in (2.3.6.3), viz.,

$$U_r N_r + V_r D_r = I_{n_i}.$$

Hence, according to Theorem 2.3.6.16.R, (N_r, D_r) is r.c.. It follows that (N_r, D_r) is a r.c.f. of H. ∎

Algorithm 2.R does more than producing a r.c.f. of H•••.

9.R Fact. Let $H \in \mathbb{R}(s)^{n_o \times n_i}$ and apply Algorithm 2.R. The result of Algorithm 2.R is that, at the end of step 3, one obtains from the submatrices of W and W^{-1} in (2.3.6.3):
(a)

(N_r, D_r) is a r.c.f. of H

with, since $WW^{-1} = I$, an associated Bezout identity:

$$U_r N_r + V_r D_r = I_{n_i}.$$

(b)

(D_ℓ, N_ℓ) is a ℓ.c.f. of H

with, since $WW^{-1} = I$, an associated Bezout identity:

$$N_\ell U_\ell + D_\ell V_\ell = I_{n_o}.$$

10 Comment. Algorithm 2.R produces simultaneously a r.c.f. and a ℓ.c.f. of H. Fact 9.R explains also the notation used in the partitions of unimodular matrices W and W^{-1} in (2.3.6.3) at the end of extraction Algorithm 2.3.6.1.R: the eight polynomial matrices N_r, D_r, U_r, V_r; N_ℓ, D_ℓ, U_ℓ, V_ℓ are data for 2 coprime fractions of the same rational matrix.

11 Proof of Fact 9.R. (a) follows by the the proof of Fact 8.R.
(b) From (2.3.6.3) with $WW^{-1} = I$, one has

12 $$N_\ell U_\ell + D_\ell V_\ell = I_{n_o}$$

13 $$N_\ell D_r = D_\ell N_r$$

where by (a) $H = N_r D_r^{-1}$.
Hence we are done if we show that

$$\det D_\ell \not\equiv 0.$$

This is done by contradiction. Assume that $\exists\ \eta \in \mathbb{R}[s]^{n_o}$, η nonzero, s.t. $\eta^T D_\ell \equiv \theta$, where θ stands for a row vector which is zero. Then by (13) and $\det D_r \not\equiv 0$, $\eta^T N_\ell \equiv \theta$. Hence by (12),

$$\theta \equiv \eta^T(N_\ell U_\ell + D_\ell V_\ell) = \eta^T \not\equiv \theta : \rightarrow\!\leftarrow.$$ ∎

14 Exercise. Use the data and results of extraction Exercise 2.3.6.6.
(a) Show that $(\bar{N}_r, \bar{D}_r)$ is a r.f. of

$$H(s) = [s + 1 \,\vdots\, -s(s - 1)^{-1}].$$

(b) Using step 3 of Algorithm 2.R, obtain a r.c.f. and a ℓ.c.f. of H with their associated Bezout identities from (2.3.6.3). ∎

We conclude this section with a result on uniqueness and the description of the Generalized Bezout Identity of any rational matrix.

17.R Theorem [Uniqueness of r.c.f.'s]. Let $H \in \mathbb{R}(s)^{n_o \times n_i}$. Then any two r.c.f.'s (N_{r1}, D_{r1}) and (N_{r2}, D_{r2}) of H are equal modulo a common unimodular right factor, or equiv.

$\exists$ a common unimodular matrix $R \in \mathbb{R}[s]^{n_i \times n_i}$, (representing e.c.o.'s on $\begin{bmatrix} D_{r1} \\ N_{r1} \end{bmatrix}$), s.t.

18 $$N_{r2} = N_{r1}R \qquad D_{r2} = D_{r1}R.$$

Proof. For i = 1, 2 (N_{ri}, D_{ri}) is r.c. Therefore, by the Bezout Identity Theorem 2.3.6.16.R for i = 1, 2 $\exists\ U_{ri}, V_{ri} \in E(\mathbb{R}[s])$ s.t.

19 $$U_{ri}N_{ri} + V_{ri}D_{ri} = I \quad \text{for } i = 1, 2.$$

Moreover, for i = 1, 2 $H = N_{ri}D_{ri}^{-1}$, whence

20 $$N_{r2} = N_{ri}D_{r1}^{-1}D_{r2}.$$

Set now

21 $$R := D_{r1}^{-1}D_{r2}.$$

From (20) and (21), (18) is established; it remains to be shown that R in (21) is unimodular. Now from (19) with i = 1, $U_{r1}N_{r2} + V_{r1}D_{r2} = R$; hence R is a polynomial matrix. From (18) $N_{r1} = N_{r2}R^{-1}$, $D_{r1} = D_{r2}R^{-1}$. Hence by (19) with i = 2, $U_{r2}N_{r1} + V_{r2}D_{r1} = R^{-1}$; hence R^{-1} is also a polynomial matrix and therefore unimodular. ∎

22 Exercise. State and prove Theorem 17.L [Uniqueness of ℓ.c.f.'s].

25.R Theorem [Generalized Bezout identity generated by a r.c.f.]. Let $H \in \mathbb{R}(s)^{n_o \times n_i}$ have a r.c.f. (N_r, D_r). Then there exist six matrices $\in E(\mathbb{R}[s])$, viz.,

$$U_r,\ V_r,\ N_\ell,\ D_\ell,\ U_\ell,\ V_\ell$$

s.t.

27 $$W\,W^{-1} := \begin{array}{c} \\ n_i \\ n_o \end{array}\!\!\overset{\quad n_i \quad\ n_i}{\left[\begin{array}{c|c} V_r & U_r \\ \hline -N_\ell & D_\ell \end{array}\right]}\overset{\quad n_i \quad\ n_o}{\left[\begin{array}{c|c} D_r & -U_\ell \\ \hline N_r & V_\ell \end{array}\right]}\!\!\begin{array}{c} \\ n_i \\ n_o \end{array} = \left[\begin{array}{c|c} I_{n_i} & 0 \\ \hline 0 & I_{n_o} \end{array}\right].$$

Moreover,

28 (D_ℓ, N_ℓ) is a ℓ.c.f. of H.

29 Comment. Equation (27) is called a generalized Bezout identity for H, e.g., [Kai.1,p.382]. It is a key tool: see below. The shaded area in (27) shows known data.

30 Proof of Theorem 25.R. Apply extraction Algorithm 2.3.6.1.R with $\bar{N}_r \leftarrow N_r$, $\bar{D}_r \leftarrow D$. Note that because of (2.3.6.4)

31 $$WM := \left[\begin{array}{c|c} V_r & U_r \\ \hline -N_\ell & D_\ell \end{array}\right]\left[\begin{array}{c} D_r \\ \hline N_r \end{array}\right] = \left[\begin{array}{c} R \\ \hline 0 \end{array}\right],$$

where, by Fact 2.3.6.10.R, R is a g.c.r.d. of (N_r, D_r). Hence by Fact 2.3.6.15.R, R is unimodular, because (N_r, D_r) is r.c. As a consequence we may set $R = I_{n_i}$ (it suffices to multiply in (31) (block ρ_1) on the left by R^{-1}: note that $R^{-1}V_rD_r + R^{-1}U_rN_r = I_{n_i}$). Now, by a proof similar to the proof of Fact 2.4.1.9.R, (D_ℓ, N_ℓ) in (31) is a ℓ.c.f. of H, whence by Theorem 2.3.6.16.R $\exists\ \tilde{U}_\ell, \tilde{V}_\ell \in E(\mathbb{R}[s])$ s.t.

$$N_\ell\tilde{U}_\ell + D_\ell\tilde{V}_\ell = I_{n_o}.$$

As a result one obtains from the above:

$$\left[\begin{array}{c|c} V_r & U_r \\ \hline -N_\ell & D_\ell \end{array}\right]\left[\begin{array}{c|c} D_r & -\tilde{U}_\ell \\ \hline N_r & \tilde{V}_\ell \end{array}\right] = \left[\begin{array}{c|c} I_{n_i} & Q \\ \hline 0 & I_{n_o} \end{array}\right],$$

where $Q := -V_r\tilde{U}_\ell + U_r\tilde{V}_\ell$. Finally, using block-column operations, viz.,

$$(\text{block } \gamma_2) \leftarrow (\text{block } \gamma_2) - (\text{block } \gamma_1)\, Q,$$

we obtain (27), defining

$$-U_\ell := \tilde{U}_\ell - D_r Q \qquad V_\ell := \tilde{V}_\ell - N_r Q.$$

We are done since (28) has already been established. ∎

32 Exercise. State Theorem 25.L [Generalized Bezout identity generated by a ℓ.c.f.].

2.4.2. Smith-McMillan Form; Relation to Coprime Fractions

The Smith-McMillan form of a rational transfer function matrix is a conceptual tool. It clarifies the relations between all coprime fractions of a rational matrix. In Sec. 2.4.4 it will be used to define poles and zeros of such a matrix.

1 Theorem [Smith-McMillan form]. Let $H \in \mathbb{R}(s)^{n_o \times n_i}$ have normal rank r. Then there exist unimodular matrices $L \in \mathbb{R}[s]^{n_o \times n_o}$ and $R \in \mathbb{R}[s]^{n_i \times n_i}$ (obtained by e.o.'s over the polynomial ring $\mathbb{R}[s]$), s.t.

2 $$H = LMR,$$

where

3 $$M := \begin{bmatrix} \operatorname{diag}[(\varepsilon_i/\psi_i)]_{i=1}^{r} & 0 \\ 0 & 0 \end{bmatrix} \begin{matrix} r \\ n_o - r \end{matrix} \in \mathbb{R}(s)^{n_o \times n_i},$$

with column blocks of widths r and $n_i - r$,

4 (ε_i, ψ_i) a pair of monic coprime polynomials for $i \in \underline{r}$,

5 $\psi_{i+1} | \psi_i$ for $i \in \underline{r-1}$,

6 $\varepsilon_i | \varepsilon_{i+1}$ for $i \in \underline{r-1}$,

7 $\psi_1 = d :=$ the monic ℓ.c.d. of all entries of H.

M is called the <u>Smith-McMillan form of H</u>. ∎

8 <u>Comment</u>. The only entries of the Smith-McMillan form, (SMM-form), that are nonzero are the diagonal entries of the r-dimensional top left block with $r = \text{rk}H$; the latter entries are rational functions.

9 <u>Proof of Theorem 1</u>. With d := the monic ℓ.c.d. of all entries of H, we can write every entry of H as

10 $$h_{ij} = n_{ij}/d,$$

where n_{ij} and d are polynomials. Therefore, by defining

11 $$N := [n_{ij}] \in \mathbb{R}[s]^{n_o \times n_i},$$

one obtains

12 $$dH = N,$$

where N is a <u>polynomial</u> matrix. Hence, according to Theorem 2.3.4.18, N has a unique Smith form S and ∃ <u>unimodular</u> matrices L and R s.t.

13 $$dH = N = LSR,$$

where

14 $$S = \begin{bmatrix} \text{diag}[\lambda_i]_{i=1}^{r} & \bigcirc \\ \bigcirc & \bigcirc \end{bmatrix} \begin{matrix} r \\ n_o - r \end{matrix}, \quad \text{column blocks: } r,\ n_i - r$$

15 $$r = \text{rk}S = \text{rk}H,$$

16 $$\lambda_i | \lambda_{i+1} \quad \text{for} \quad i \in \underline{r-1}.$$

Finally, after dividing (13) by d:

17 $$H = LMR,$$

where

18 $$M = S\,d^{-1} = \left[\begin{array}{c|c} \operatorname{diag}[(\varepsilon_i/\psi_i)]_{i=1}^{r} & \bigcirc \\ \hline \bigcirc & \bigcirc \end{array}\right]\begin{matrix} r \\ n_o - r \end{matrix} ,$$
$$\qquad\qquad\qquad\quad r \qquad\qquad\qquad n_i - r$$

because after canceling common factors,

19 $$(\lambda_i/d) =: (\varepsilon_i/\psi_i) \in \mathbb{R}(s) \quad \text{for} \quad i \in \underline{r},$$

with

20 (ε_i, ψ_i) a pair of monic coprime polynomials for $i \in \underline{r}$.

Hence we have proved (2)-(4).

Division properties (5) and (6) are consequences of (16), (19), and (20). Indeed, $\lambda_{i+1}/\lambda_i = (\varepsilon_{i+1}\ \psi_i)/(\varepsilon_i\ \psi_{i+1})$ is a polynomial where $(\varepsilon_i,\ \psi_i)$ and $(\varepsilon_{i+1},\ \psi_{i+1})$ are coprime polynomials; therefore, $\varepsilon_i | \varepsilon_{i+1}$ and $\psi_{i+1} | \psi_i$.

Property (7) is established by contradiction. Assume therefore that $\psi_1 \neq d$:= the monic least common denominator of all entries of H. Then, because $(\varepsilon_1/\psi_1) = (\lambda_1/d)$ with $(\varepsilon_1,\ \psi_1)$ coprime, the polynomials λ_1 and d must have a common factor (one gets $(\varepsilon_1,\ \psi_1)$ coprime by canceling common factors in (λ_1/d)). Now, by the Smith form Theorem 2.3.4.18, $\lambda_1 = \Delta_1$ = the monic g.c.d. of <u>all</u> entries of polynomial matrix N (see (10)-(16)). Hence d and <u>all</u> entries of N have a common factor. This contradicts the fact that d is the <u>least</u> common denominator of every entry of H. ∎

Recall that by Theorems 2.4.1.17.R and 2.4.1.17.L all r.c.f.'s (ℓ.c.f.'s) of a given rational matrix $H \in \mathbb{R}(s)^{n_o \times n_i}$ are equal modulo a common right (resp. left) unimodular factor. A convenient way to relate arbitrary coprime fractions (right and/or left) is by using the SMM-form. Therefore, consider $H \in \mathbb{R}(s)^{n_o \times n_i}$ and its SMM-form as described in Theorem 1 by (2)-(7).

Define now the following polynomial matrices based on (2)-(7):

25 $$\boldsymbol{\varepsilon} := \left[\begin{array}{c|c} \text{diag}[\varepsilon_i]_{i=1}^r & \bigcirc \\ \hline \bigcirc & \bigcirc \end{array}\right] \begin{array}{l} r \\ n_o - r \end{array}, \qquad \text{columns: } r,\ n_i - r$$

26 $$\boldsymbol{\Psi}_r := \left[\begin{array}{c|c} \text{diag}[\psi_i]_{i=1}^r & \bigcirc \\ \hline \bigcirc & I \end{array}\right] \begin{array}{l} r \\ n_i - r \end{array}, \qquad \text{columns: } r,\ n_i - r$$

27 $$\boldsymbol{\Psi}_\ell := \left[\begin{array}{c|c} \text{diag}[\psi_i]_{i=1}^r & \bigcirc \\ \hline \bigcirc & I \end{array}\right] \begin{array}{l} r \\ n_o - r \end{array}, \qquad \text{columns: } r,\ n_o - r$$

28 $$N_r := L\boldsymbol{\varepsilon}, \quad D_r := R^{-1}\boldsymbol{\Psi}_r,$$

29 $$D_\ell := \boldsymbol{\Psi}_\ell L^{-1}, \quad N_\ell := \boldsymbol{\varepsilon} R.$$

Note that (a) the <u>dimensions</u> of $\boldsymbol{\Psi}_r$ and $\boldsymbol{\Psi}_\ell$ have been <u>adapted</u> to match n_i := the # of inputs and n_o := the # of outputs, and (b) that $\boldsymbol{\Psi}_r$ and $\boldsymbol{\Psi}_\ell$ are in <u>reverse Smith form</u> $(\psi_i | \psi_{i-1})$.

Notice now that by Theorem 1 and the definitions above

30 $$H = N_r D_r^{-1} = D_\ell^{-1} N_\ell,$$

where at every $s \in \mathbb{C}$

31 $$\text{rk}\left[\begin{array}{c} D_r \\ \hline N_r \end{array}\right] = \text{rk}\underbrace{\left[\begin{array}{c|c} R^{-1} & \bigcirc \\ \hline \bigcirc & L \end{array}\right]}_{\text{unimodular}}\left[\begin{array}{c} \boldsymbol{\Psi}_r \\ \hline \boldsymbol{\varepsilon} \end{array}\right] = \text{rk}\left[\begin{array}{c} \boldsymbol{\Psi}_r \\ \hline \boldsymbol{\varepsilon} \end{array}\right] = n_i,$$

and similarly at every $s \in \mathbb{C}$

32 $$\mathrm{rk}[D_\ell \,\vdots\, N_\ell] = \mathrm{rk}[\Psi_\ell \ \ \mathcal{E}] = n_o .$$

From (30) we have that (N_r, D_r) $((D_\ell, N_\ell))$ is a r.f. (resp. ℓ.f.) of $H \in \mathbb{R}(s)^{n_o \times n_i}$. By rank test Theorems 2.3.6.20.R and 2.3.6.20.L it follows that the pairs (N_r, D_r) and (D_ℓ, N_ℓ) are coprime. Hence

33 <u>Fact</u>. The SMM-form generates through matrices (25)-(29) a r.c.f. (N_r, D_r) and a ℓ.c.f. (D_ℓ, N_ℓ) for any $H \in \mathbb{R}(s)^{n_o \times n_i}$. ■

We shall now relate arbitrary coprime fractions of $H \in \mathbb{R}(s)^{n_o \times n_i}$ using the notion of equivalence (see Sec. 2.3.2), and uniqueness Theorems 2.4.1.17.R and 2.4.1.17.L.

36 <u>Theorem</u> [Numerators of coprime fractions]. Let $H \in \mathbb{R}(s)^{n_o \times n_i}$. Then all the numerators of any coprime fraction of H (right and/or left) are <u>equivalent</u> and have the same Smith form $\mathcal{E}$ given by (25).

37 <u>Comment</u>. Numerators can be obtained from each other by e.o.'s: they have the same normal rank and the same local rank at every $s \in \mathbb{C}$.

38 <u>Proof of Theorem 36</u>. By the uniqueness Theorems 2.4.1.17.R and 2.4.1.17.L we have only to investigate the relation between numerators N_r and N_ℓ of Fact 33. Now, by (28)-(29) with L and R unimodular, $N_r \sim \mathcal{E} \sim N_\ell$. Therefore, equivalence follows with as unique Smith form, matrix $\mathcal{E}$ given by (25): notice that $\varepsilon_i | \varepsilon_{i+1}$. ■

41 <u>Theorem</u> [Denominators of coprime fractions]. Let $H \in \mathbb{R}(s)^{n_o \times n_i}$. Then all the denominators of any coprime fraction of H (right and/or left) have the <u>same nonunity invariant polynomials</u>. Hence, in particular, their <u>determinants are equivalent</u> (equiv. equal modulo a nonzero constant). Moreover, if $n_o = n_i$, (equiv. <u>H is square</u>), then these denominators are <u>equivalent</u>.

42 Comment. In the nonsquare case there is no equivalence because right and left denominators have different dimensions.

43 Exercise. Prove Theorem 41. (Hint: Use the uniqueness Theorems 2.4.1.17.R and 2.1.4.17.L and observe that (28)-(29) $D_\ell \sim \Psi_\ell$, $D_r \sim \Psi_r$, while in (26)-(27) Ψ_ℓ and Ψ_r have the same nonunity invariant polynomials.)

44 Exercise. Let $H \in \mathbb{R}(s)^{n_o \times n_i}$, where $n_o > n_i$. Let (N_r, D_r) be a r.c.f. and (D_ℓ, N_ℓ) be a ℓ.c.f. of H. Show that

$$D_\ell \sim \left[\begin{array}{c|c} D_r & 0 \\ \hline 0 & I_{n_o - n_i} \end{array}\right].$$

2.4.3. Proper Transfer Function Matrices

Most physical models have low-pass frequency characteristics, hence finite gain as the frequency is increased···. For linear time-invariant systems described by a transfer function this means that the latter has to be bounded at infinity.

1 Definition. Let $H \in \mathbb{R}(s)^{n_o \times n_i}$. Then we say that H is proper (strictly proper) iff $\lim_{s\to\infty} H(s) = H(\infty) \in \mathbb{C}^{n_o \times n_i}$ ($\lim_{s\to\infty} H(s) = 0$, resp.). This is denoted by $H \in \mathbb{R}_p(s)^{n_o \times n_i}$ ($H \in \mathbb{R}_{p,o}(s)^{n_o \times n_i}$, resp.), where $\mathbb{R}_p(s)$ ($\mathbb{R}_{p,o}(s)$) is the ring of proper (strictly proper) rational functions with coefficients in $\mathbb{R}$. ∎

2 Exercise. Show that $\mathbb{R}_p(s)$ and $\mathbb{R}_{p,o}(s)$ are subrings of the field $\mathbb{R}(s)$.

3 Definition. Let $m \in \mathbb{R}[s]^n$ be a polynomial vector. We define the degree of m, denoted by $\partial[m]$, to be the highest degree of all entries of the vector. If the vector m is row i or column j of the polynomial matrix $M \in \mathbb{R}[s]^{n_o \times n_i}$, then their degrees are called ith row-degree or jth column-degree and denoted by $\partial_{ri}[M]$, $\partial_{cj}[M]$ resp. Note that for $m \in \mathbb{R}[s]^n$, $\partial[m] = -\infty$ iff m is the zero vector.

A rational function is proper if and only if the degree of the denominator is at least the degree of the numerator. In the matrix case this is only necessary column-wise or row-wise. (However, see 25.R below.)

4.R Fact. Let $H \in \mathbb{R}_p(s)^{n_o \times n_i}$ ($\in \mathbb{R}_{p,o}(s)^{n_o \times n_i}$) have a r.f. (N_r, D_r); then

5 $\quad \forall j \in \underline{n}_i \quad \partial_{cj}[N_r] \leq \partial_{cj}[D_r]$

6 $\quad (\forall j \in \underline{n}_i \quad \partial_{cj}[N_r] < \partial_{cj}[D_r]$, resp.).

Proof. We prove the proper case and drop subscripts r. Note that N = HD with $N = [n_{ij}]$, $D = [D_{ij}]$ reads

$$n_{ij} = \sum_{k=1}^{n_i} h_{ik} d_{kj}.$$

Let now $k_j := \partial_{cj}[D]$; then taking limits we have

$$\lim_{s\to\infty} n_{ij}(s)\, s^{-k_j} = \sum_{k=1}^{n_i} (\lim_{s\to\infty} h_{ik}(s))\, (\lim_{s\to\infty} d_{kj}(s)\, s^{-k_j}),$$

where all limits on the RHS exist and are finite. Hence the limit on the LHS must exist and be finite. As a consequence, $\forall i \in \underline{n}_o, \partial[n_{ij}] \leq k_j := \partial_{cj}[D]$. Hence taking the maximum over i in the LHS, (5) follows. ∎

7 Exercise. State and prove Fact 4.L. (Let $H \in \mathbb{R}_p(s)^{n_o \times n_i}$ have a ℓ.f. (D_ℓ, N_ℓ) then $\forall i \in \underline{n}_o\ \partial_{ri}[N_\ell] \leq \cdots$.)

8 Remark. the converse of Fact 4.R is not true in general. For example, with

$$N_r(s) = [s \mid 1], \quad D_r(s) = \begin{bmatrix} 1 & 1 \\ s & s^2 \end{bmatrix}$$

condition (5) is satisfied, but $H(s) = \begin{bmatrix} \frac{s^3-s}{s^2-s} & \frac{1-s}{s^2-s} \end{bmatrix} \notin \mathbb{R}_p(s)^{1\times 2}$.

In the analysis below a condition is added to make the converse true....

11 Definition. Let $D \in \mathbb{R}[s]^{n\times n}$ be a nonsingular polynomial matrix. We say that D is column-reduced (c.r.) (resp. row-reduced (r.r.)) iff

$\partial[\det D] = \sum_{j=1}^{n} \partial_{cj}[D]$ ($\partial[\det D] = \sum_{i=1}^{n} \partial_{ri}[D]$, resp.).

12 Exercise. Show that one always has

13 $$\partial[\det D] \leq \sum_{j=1}^{n} \partial_{cj}[D].$$

(Hint: Set $k_j := \partial_{cj}[D]$ and consider $\lim_{s\to\infty} (\det D(s)\, (s^{-\sum_{j=1}^{n} k_j}))$.) ∎

14 Remark. Exercise 12 suggests that column-reducedness can be achieved by using e.c.o.'s to successively reduce the individual column-degrees until column-reducedness is achieved. For example, the e.c.o. $\gamma_2 \leftarrow \gamma_2 - s\gamma_1$ will transform

$$D(s) = \begin{bmatrix} 1 & 1 \\ s & s^2 \end{bmatrix} \text{ into } \begin{bmatrix} 1 & 1 - s \\ s & 0 \end{bmatrix}, \text{ which is c.r.}$$

See also, e.g., [Kai.1, Example 6.3-2]. As a consequence, if $H \in \mathbb{R}(s)^{n_o \times n_i}$ has a r.f. $(\bar{N}_r, \bar{D}_r)$, then by e.c.o.'s on matrix $\begin{bmatrix} \bar{D}_r \\ \bar{N}_r \end{bmatrix}$, H will have a r.f. (N_r, D_r) where D_r is c.r.; moreover, if $(\bar{N}_r, \bar{D}_r)$ is a r.c.f., then (N_r, D_r) remains a r.c.f. Similar statements can be made for row-reducedness. Hence without loss of generality we may assume that the denominator of a (coprime) fraction is column- or row-reduced.

15 Highest Column-degree Coefficient Matrix. For any nonsingular polynomial matrix $D \in \mathbb{R}[s]^{n\times n}$, let

16 $$k_j := \partial_{cj}[D] \quad \text{for } j \in \underline{n};$$

then

17 $$\lim_{s\to\infty} D(s)\, (\text{diag}[s^{-k_j}]_{j=1}^{n}) =: D_h \in \mathbb{R}^{n\times n},$$

where $D_h \in \mathbb{R}^{n\times n}$ is the highest column-degree coefficient matrix of D (representing the coefficients of s^{k_j} of each entry d_{ij} of D). Moreover, since the determinant is continuous in its arguments

18 $$\lim_{s\to\infty} (\det D(s)\, (s^{-\sum_{j=1}^{n} k_j})) = \det D_h \in \mathbb{R}\,.$$ ■

We have then, by Definition 11 and (18),

21.R <u>Fact</u>. Let $D \in \mathbb{R}[s]^{n\times n}$ be a nonsingular polynomial matrix with column-degrees k_j and highest column-degree coefficient matrix $D_h \in \mathbb{R}^{n\times n}$.

U.t.c.

D is column-reduced (c.r.)

⇔

D_h is nonsingular

⇔

$$\lim_{s\to\infty} D(s)\, (\mathrm{diag}[s^{-k_j}]_{j=1}^{n}) = D_h, \text{ with } D_h \text{ nonsingular.}$$ ■

22 <u>Exercise</u>. State Fact 21.L (row-reducedness, row-degrees, highest row-degree coefficient matrix).

Fact 4.R can now be refined to an equivalence.

25.R <u>Theorem</u>. Let $H \in \mathbb{R}(s)^{n_o\times n_i}$ admit a r.f. (N_r, D_r) with D_r c.r..

U.t.c.

$$H \in \mathbb{R}_p(s)^{n_o\times n_i}, \ (H \in \mathbb{R}_{p,o}(s)^{n_o\times n_i})$$

⇔

26 $$\forall j \in \underline{n}_i \qquad \partial_{cj}[N_r] \le \partial_{cj}[D_r]$$

27 $$(\forall j \in \underline{n}_i \qquad \partial_{cj}[N_r] < \partial_{cj}[D_r], \text{ resp.}).$$

28 <u>Comment</u>. Modulo a column-reduced denominator the characterization of a proper right fraction is a straightforward extension of the scalar case.

29 <u>Proof of Theorem 25.R</u>. ⇒ : follows by Fact 4.R.

⇐ : Drop subscripts r. Then, using the column-degrees k_j of D as given by (16), one obtains

$$H(s) = (N(s)\,\mathrm{diag}[s^{-k_j}])\,(D(s)\,\mathrm{diag}[s^{-k_j}])^{-1},$$

where

(a) $\lim_{s\to\infty} D(s)\,\mathrm{diag}[s^{-k_j}] = D_h$, with D_h nonsingular since D is c.r. and by Fact 21.R;

(b) $\lim_{s\to\infty} N(s)\,\mathrm{diag}[s^{-k_j}] =: N_h \in \mathbb{R}^{n_o\times n_i}$ (= 0, resp.) because of (26), ((27) resp.).

Hence $\lim_{s\to\infty} H(s) = N_h D_h^{-1}$ (= 0, resp.). ∎

Two other properties are important; we state the first.

32.R Theorem [Uniqueness of column-degrees], e.g., [Kai.1, Lemma 6.3.14]. Let $D \in \mathbb{R}[s]^{n\times n}$ and $\bar{D} \in \mathbb{R}[s]^{n\times n}$ be two column-reduced nonsingular polynomial matrices with column-degrees arranged in the same order (increasing or decreasing). U.t.c., if D and $\bar{D}$ are right equivalent, then their column-degrees are identical.

33.R Comment. Theorem 32.R shows that if $H \in \mathbb{R}(s)^{n_o\times n_i}$ has a r.c.f. (N_r, D_r) with D_r column-reduced and column-degrees in increasing or decreasing order, then these column-degrees are a property of H: all r.c.f.'s with the properties mentioned above are right equivalent and therefore display the same column-degrees.

34 Exercise. State Theorem 32.L. [Uniqueness of row-degrees]. What about Comment 33.L?

37.R Theorem [Division Theorem], e.g., [Kai.1, Theorem 6.3.15]. Let $D_r \in \mathbb{R}[s]^{n_i\times n_i}$ be a nonsingular polynomial matrix. Then for any polynomial matrix $N_r \in \mathbb{R}[s]^{n_o\times n_i}$ there exist unique polynomial matrices $Q_r \in \mathbb{R}[s]^{n_o\times n_i}$ and $R_r \in \mathbb{R}[s]^{n_o\times n_i}$ such that

38 $$N_r = Q_r D_r + R_r \quad \text{with} \quad R_r D_r^{-1} \in \mathbb{R}_{p,o}(s)^{n_o\times n_i}.$$

Moreover, if D_r is also c.r., then the uniqueness of Q_r and R_r will also be ensured iff

39 $$\partial_{cj}[R_r] < \partial_{cj}[D_r] \quad \forall j \in \underline{n}_i.$$

Proof. Observe that $H := N_r D_r^{-1} \in \mathbb{R}(s)^{n_o \times n_i}$; hence

$$\forall i,j \quad h_{ij} := n_{ij}/d_{ij} \in \mathbb{R}(s),$$

where (n_{ij}, d_{ij}) is a coprime polynomial pair. Therefore, by the Euclid algorithm $\forall i,j \ \exists!$ polynomials q_{ij}, r_{ij} s.t.

$$n_{ij} = d_{ij}q_{ij} + r_{ij} \quad \text{with} \quad r_{ij}/d_{ij} \in \mathbb{R}_{p,o}(s).$$

Hence

$$\forall i,j \quad h_{ij} = q_{ij} + r_{ij}/d_{ij},$$

s.t. with

$$Q_r := [q_{ij}] \in \mathbb{R}[s]^{n_o \times n_i} \text{ and } H_{sp} := [r_{ij}/d_{ij}] \in \mathbb{R}_{p,o}(s)^{n_o \times n_i}$$

$$N_r D_r^{-1} = H = Q_r + H_{sp}.$$

Hence with

$$R_r := N_r - Q_r D_r \in \mathbb{R}[s]^{n_o \times n_i}.$$

(38) follows since $R_r D_r^{-1} = H_{sp} \in \mathbb{R}_{sp}(s)^{n_o \times n_i}$. By Theorem 25.R the latter holds. The uniqueness of the pair (Q_r, R_r) holds because, if $(\bar{Q}_r, \bar{R}_r)$ is another pair such that (38) holds, then, from $N_r = Q_r D_r + R_r = \bar{Q}_r D_r + \bar{R}_r$,

$$Q_r - \bar{Q}_r = (\bar{R}_r - R_r)\, D_r^{-1},$$

where the LHS $\in E(\mathbb{R}[s])$ and the RHS $\in E(\mathbb{R}_{p,o}(s))$. Now $\mathbb{R}[s] \cap \mathbb{R}_{p,o}(s) = \{0\}$. Hence $Q_r = \bar{Q}_r$ and $R_r = \bar{R}_r$. ∎

40 Exercise. State Theorem 37.L [Division on the left]. As a final comment it should be stressed that division can be performed by an algorithm.

41 Exercise. (a) Study the algorithmic division on the left of a polynomial vector by a nonsingular polynomial matrix in Appendix C.

(b) Let

$$N_\ell(s) := \left[\begin{array}{c|c} s^4 & s^3 \\ \hline s^2 & s^3 \end{array}\right], \quad D_\ell(s) := \left[\begin{array}{c|c} s^3 + 4s^2 + 5s + 2 & s + 2 \\ \hline s + 1 & s^2 + 4s + 4 \end{array}\right].$$

Show that the quotient Q_ℓ and remainder R_ℓ of the division on the left of N_ℓ by D_ℓ is given by

$$Q_\ell(s) := \left[\begin{array}{c|c} s - 4 & 1 \\ \hline 0 & s - 4 \end{array}\right], \quad R_\ell(s) = \left[\begin{array}{c|c} 11s^2 + 18s + 8 & -5s^2 - 3s + 6 \\ \hline 3s + 4 & 11s + 15 \end{array}\right].$$

(Hint for (b): The division on the left of N_ℓ by D_ℓ is obtained by successively dividing the columns of N_ℓ by D_ℓ.)

2.4.4. Poles and Zeros

1. Definitions. Let $H \in \mathbb{R}(s)^{n_o \times n_i}$ and consider its SMM-form $M \in \mathbb{R}(s)^{n_o \times n_i}$ (Theorem 2.3.2.1), given by

$$M = \left[\begin{array}{c|c} \operatorname{diag}[(\varepsilon_i/\psi_i)]_{i=1}^{r} & 0 \\ \hline 0 & 0 \end{array}\right] \begin{array}{l} r \\ n_o - r \end{array}, \qquad \begin{array}{cc} r & n_i - r \end{array}$$

where, for $i \in \underline{r}$, ε_i and ψ_i are monic coprime polynomials having properties (2.4.2.5)-(2.4.2.7) and where r is the normal rank of H.

We call pole of H any root of any denominator polynomial ψ_i, $i \in \underline{r}$, of the SMM-form of H. The set of poles of H is denoted by $P[H]$.

We call zero of H any root of any numerator polynomial ε_i, $i \in \underline{r}$, of the SMM-form of H. The set of zeros of H is denoted by $Z[H]$. ■

2 Comment. Note that a pole of H may also be a zero of H, e.g., let $M = \operatorname{diag}\{\frac{1}{s+1}, 1, (s+1)\}$. So for the multivariable case, poles and zeros may coincide: this is due to the fact that they have value in $\mathbb{C}$ and "place in the matrix."

3 Fact. Let $H \in \mathbb{R}(s)^{n_o \times n_i}$ with $H = [h_{ij}]_{i \in \underline{n}_o, j \in \underline{n}_i}$. Then

(a) $p \in P[H] \Leftrightarrow \exists\, h_{ij}$ s.t. $p \in P[h_{ij}]$;

(b) $p \notin P[H] \Leftrightarrow$ the map $s \mapsto H(s)$ in bounded in $N(p)$, where $N(p)$ denotes a sufficiently small neighborhood of $p \in \mathbb{C}$.

4 Comments. (a) means that at least one entry of H has a pole at p.

(b) Since the map $s \mapsto H(s)$ is meromorphic in s (equiv. has $\forall p \in \mathbb{C}$ a Laurent expansion in some $N(p)$ which stops at most at a negative power of $(s-p)$), the map $s \mapsto H(s)$ is bounded in some $N(p)$ iff the map $s \mapsto H(s)$ is analytic at p.

5 Proof of Fact 3. (a) Recall Definition 1 and the fact that $\psi_i | \psi_{i-1}$; consequently, $p \in P[H]$ if and only if $\psi_1(p) = d(p) = 0$, where d is the ℓ.c.d. of the entries of H. Hence $p \in P[H]$ iff $\exists\, h_{ij}$ s.t. $p \in P[h_{ij}]$.

(b) By the negation of (a):

$$p \notin P[H] \Leftrightarrow \forall i,j,\ p \notin P[h_{ij}]$$

$\Leftrightarrow \forall i,j$, map $s \mapsto h_{ij}(s)$ is bounded in some $N_{ij}(p)$

$\Leftrightarrow \forall i,j$, map $s \mapsto H(s)$ is bounded in some $N(p)$

(take the intersection of the $N_{ij}(p)$'s above). ∎

We shall now characterize poles and zeros in terms of any coprime fraction of a rational transfer function matrix. Recall therefore the developments of Sec. 2.4.2 on the SMM-form and its relation to coprime fractions. To avoid unnecessary redundancy in the transfer function system description, we shall also assume that the normal rank of $H \in \mathbb{R}(s)^{n_o \times n_i}$ is $r = \min(n_o, n_i)$, (otherwise, there are "trivial inputs or outputs").

6 Theorem [Poles and zeros of coprime fractions]. Let $H \in \mathbb{R}(s)^{n_o \times n_i}$ have normal rank $r = \min(n_o, n_i)$. Let (N_r, D_r), resp. (D_ℓ, N_ℓ), be any r.c.f. or ℓ.c.f. of H.

U.t.c.

(a)

$$p \in P[H] \Leftrightarrow \det D_r(p) = 0 \Leftrightarrow \det D_\ell(p) = 0$$

(b)

$z \in Z[H]$

$\Leftrightarrow$

$$\mathrm{rk}[N_r(z)] < r = \min(n_o, n_i)$$

$\Leftrightarrow$

$$\mathrm{rk}[N_\ell(z)] < r = \min(n_o, n_i).$$

7 Proof of Theorem 6. (a) According to denominator Theorem 2.4.2.41 and Fact 2.4.2.33 using (2.4.2.25)-(2.4.2.29) we have

$$\det D_\ell \sim \det D_r \sim \det \Psi_r \sim \det \Psi_\ell.$$

Now according to Definitions 1 we have

$$p \in P[H] \;\Leftrightarrow\; \det \Psi_r(p) = 0 \;\Leftrightarrow\; \det \Psi_\ell(p).$$

Hence assertion (a) is true.

(b) According to Numerator Theorem 2.4.2.36 and Fact 2.4.2.33, using (2.4.2.25)-(2.4.2.29) we have $N_\ell \sim N_r \sim \mathcal{E}$. Now according to Definitions 1 we have $z \in Z[H] \Leftrightarrow \mathrm{rk}[\mathcal{E}(z)] < r$. Hence assertion (b) is true. ∎

An immediate consequence of Theorem 6 is that zeros are poles of the inverse transfer function matrix, and vice versa.

8 Fact. Let $H \in \mathbb{R}(s)^{n\times n}$ be nonsingular. Then

$$z \in Z[H] \;\Leftrightarrow\; p \in P[H^{-1}];\; p \in P[H] \;\Leftrightarrow\; p \in Z[H^{-1}].$$

9 Exercise. Prove Fact 8. (Hint: If (N_r, D_r) is a r.c.f. of H, then (D_r, N_r) is a r.c.f. of ⋯.)

Recall now that a state space system description (SSD) is a quadruple [A, B, C, D] of real matrices describing $\dot{x} = Ax + Bu$, $y = Cx + Du$ with A, B, C, $D \in \mathbb{R}^{n\times n}$, $\mathbb{R}^{n\times n_i}$, $\mathbb{R}^{n_o\times n}$, $\mathbb{R}^{n_o\times n_i}$, resp.. Any SSD [A, B, C, D] defines a transfer function $H(s) = C(sI - A)^{-1}B + D \in \mathbb{R}_p(s)^{n_o\times n_i}$, where without loss of generality $\mathrm{rk}[C \,\vdots\, D] = n_o$ and $\mathrm{rk}\begin{bmatrix} B \\ \hdashline D \end{bmatrix} = n_i$ (otherwise, there are "trivial outputs or inputs"). A minimal realization of $H(s) \in \mathbb{R}_p(s)^{n_o\times n_i}$ is a SSD [A, B, C, D] s.t. A has minimal dimension. We have then

12 Theorem [Poles and zeros of minimal realizations]. Let $H \in \mathbb{R}_p(s)^{n_o\times n_i}$

and let [A, B, C, D] be any minimal realization of order $n = \partial[\det(sI - A)]$, with $\mathrm{rk}\begin{bmatrix} B \\ \hline D \end{bmatrix} = n_i$ and $\mathrm{rk}[C \mid D] = n_o$.

U.t.c.

(a)

13 $\quad p \in P[H] \;\Leftrightarrow\; \det(pI - A) = 0;$

(b) if

14 $$P(s) := \begin{array}{c} \\ n \\ n_o \end{array}\!\!\begin{array}{c} \begin{array}{cc} n & n_i \end{array} \\ \left[\begin{array}{c|c} sI - A & B \\ \hline -C & D \end{array}\right] \end{array}$$

denotes the system matrix [Ros.1], then

15 $$z \in Z[H] \;\Leftrightarrow\; \mathrm{rk}[P(z)] < n + \min(n_o, n_i). \qquad ■$$

16 Proof of Theorem 12. Since (A, B) and (C, A) are controllable resp. observable, $(sI - A)^{-1}B$ is a ℓ.c.f. and $C(sI - A)^{-1}$ is a r.c.f.. Therefore, by Theorems 2.4.1.25L and 2.4.1.25R we have two generalized Bezout identities, viz.,

17 $$\left[\begin{array}{c|c} \bar{V}_r & \bar{U}_r \\ \hline -B & sI - A \end{array}\right]\left[\begin{array}{c|c} \bar{D}_r & -\bar{U}_\ell \\ \hline \bar{N}_r & \bar{V}_\ell \end{array}\right] = \left[\begin{array}{c|c} I & 0 \\ \hline 0 & I \end{array}\right],$$

18 $$\left[\begin{array}{c|c} \tilde{V}_r & \tilde{U}_r \\ \hline -\tilde{N}_\ell & \tilde{D}_\ell \end{array}\right]\left[\begin{array}{c|c} sI - A & -\tilde{U}_\ell \\ \hline C & \tilde{V}_\ell \end{array}\right] = \left[\begin{array}{c|c} I & 0 \\ \hline 0 & I \end{array}\right],$$

where

19 $$(\bar{N}_r, \bar{D}_r) \text{ is a r.c.f. of } (sI - A)^{-1}B$$

and where by Theorem 2.4.2.41

20 $$\det(sI - A) \sim \det \bar{D}_r.$$

Claim 1: $(C\bar{N}_r, \bar{D}_r)$ is a r.c.f. of $C(sI - A)^{-1}B$.

By (19) it is sufficient to show that $(C\bar{N}_r, \bar{D}_r)$ is r.c. Now from the generalized Bezout identities (17)-(18) one has

21 $$\tilde{V}_r(sI - A) + \tilde{U}_r C = I,$$

22 $$\bar{V}_r\bar{D}_r + \bar{U}_r\bar{N}_r = I.$$

On multiplying (21) on the right by $(sI - A)^{-1}B$ and using (19), we get successively

$$\tilde{V}_r B + \tilde{U}_r C(sI - A)^{-1}B = (sI - A)^{-1}B,$$

$$\tilde{V}_r B + \tilde{U}_r C\bar{N}_r\bar{D}_r^{-1} = \bar{N}_r\bar{D}_r^{-1},$$

$$\tilde{V}_r B\bar{D}_r + \tilde{U}_r C\bar{N}_r = \bar{N}_r.$$

On substituting this expression for $\bar{N}_r$ in (22), we get finally

$$(\bar{V}_r + \bar{U}_r\tilde{V}_r B)\bar{D}_r + (\bar{U}_r\tilde{U}_r)C\bar{N}_r = I.$$

This is a Bezout identity for $(C\bar{N}_r, \bar{D}_r)$, which is therefore r.c. ∎

Claim 2: $(C\bar{N}_r + D\bar{D}_r, \bar{D}_r)$ is a r.c.f. of H.

The proof of this claim is left as an exercise.

Claim 3: The system matrix $P(\cdot)$ defined in (14) is equivalent to

$$\left[\begin{array}{c|c} I_n & 0 \\ \hline 0 & C\bar{N}_r + D\bar{D}_r \end{array}\right].$$

Indeed, using (17) (where the matrices on the LHS are unimodular), and starting by the product of $P(\cdot)$ and a unimodular matrix we obtain successively

$$\left[\begin{array}{c|c} sI - A & B \\ \hline -C & D \end{array}\right]\left[\begin{array}{c|c} \bar{V}_\ell & \bar{N}_r \\ \hline \bar{U}_\ell & -\bar{D}_r \end{array}\right] = \left[\begin{array}{c|c} I_n & 0 \\ \hline -C\bar{V}_\ell + D\bar{U}_\ell & -C\bar{N}_r - D\bar{D}_r \end{array}\right]$$

$$\sim \left[\begin{array}{c|c} I_n & 0 \\ \hline 0 & C\bar{N}_r + D\bar{D}_r \end{array}\right].$$

Thus claim 3 follows. ■

Conclusion. Using Claim 2, Theorem 6, (20), and Claim 3:

(a) $p \in P[H] \Leftrightarrow \det \bar{D}_r(p) = 0 \Leftrightarrow \det(pI - A) = 0$

(b) $z \in Z[H] \Leftrightarrow \text{rk}[C\bar{N}_r + D\bar{D}_r](z) < \min(n_o, n_i)$

$\Leftrightarrow \text{rk}[P(z)] < n + \min(n_o, n_i)$.

End of Proof

The following properties of zeros are consequences of Theorem 12.

25 Fact. Let [A, B, C, 0] be any minimal realization of $H \in \mathbb{R}_{p,o}(s)^{n_o \times n_i}$ with $\text{rk}[B] = n_i$ and $\text{rk}[C] = n_o$. Then using characterization (15), the zeros of H are invariant under

(a) algebraic equivalence ($A \leftarrow TAT^{-1}$, $B \leftarrow TB$, $C \leftarrow CT^{-1}$, $\det T \neq 0$),

(b) constant output feedback ($A \leftarrow A + BKC$, $B \leftarrow B$, $C \leftarrow C$). ■

26 Exercise. Prove Fact 25 (Hint: Note that minimality is conserved because algebraic equivalence and constant output feedback upset neither controllability nor observability, e.g., [Che.1]; use also (15).)

To motivate the next theorem consider the following:

27 Exercise. Let

$$h(s) := ((s - 1)(s - 2))/((s + 1)(s + 2)(s + 3)) \in \mathbb{R}_{p,o}(s)$$

and consider the controllable, resp. observable, canonical realizations [A, b, c, o] of h, e.g., [Che.1, Theorems 7-1 and 7-2].

(a) Find a state feedback vector $k \in \mathbb{R}^{1\times n}$, representing gains from each state to the input of the controllable realization, such that the numerator of the resulting transfer function becomes a nonzero constant (equiv. all the zeros of h(s) have been canceled).

(b) Find a feedback vector $k \in \mathbb{R}^{n\times 1}$, representing gains from the output to each integrator input of the observable realization, such that the numerator of the resulting transfer function becomes a nonzero constant. ∎

Exercise 27 shows that for the scalar case pole-zero cancellation is associated with (a) suitable constant state to input feedback ($A \leftarrow A + BK$, $B \leftarrow B$, $C \leftarrow C$) and (b) suitable constant output to integrator-input feedback ($A \leftarrow A + KC$, $B \leftarrow B$, $C \leftarrow C$). Note that in the literature (a) constant state to input feedback is known as state feedback, e.g., [Che.1, Sec. 7-3], and (b) constant output to integrator-input feedback is related to state reconstruction by observers, e.g., [Che.1, Sec. 7-4]. This leads to the following characterization of zeros in the multivariable case.

28 Theorem [Pole-zero cancellation under constant state to input feedback and constant output to integrator-input feedback]. Let $H \in \mathbb{R}_{p,o}(s)^{n_o\times n_i}$ with a minimal realization $[A, B, C, 0]$ and $rk(B) = n_i$, $rk(C) = n_o$.

(a) Let $n_i \le n_o$; then

29
$$z \in Z[H] \Leftrightarrow \begin{cases} \exists \text{ a nonzero vector } x_0 \in \mathbb{C}^n, \ \exists \text{ a matrix } K \in \mathbb{C}^{n_i\times n} \\ \text{s.t. } (A + BK)x_0 = zx_o,\ Cx_o = \theta \end{cases}$$

or equiv., for any zero z, there is a constant state to input feedback matrix $K \in \mathbb{C}^{n_i\times n}$ such that for the resulting system $[A + BK, B, C, 0]$ the zero z is a system eigenvalue which is unobservable.

(b) Let $n_o \le n_i$; then

30
$$z \in Z[H] \Leftrightarrow \begin{cases} \exists \text{ a nonzero vector } x_0 \in \mathbb{C}^n, \ \exists \text{ a matrix } K \in \mathbb{C}^{n\times n_o} \\ \text{s.t. } (A + KC)^T x_0 = zx_o,\ B^T x_o = \theta, \end{cases}$$

or equiv., for any zero z, there is a constant output to integrator-input matrix $K \in \mathbb{C}^{n\times n_o}$ such that for the resulting system $[A + KC, B, C, 0]$ the zero z is a system eigenvalue which is uncontrollable.

31 Comment. Note, e.g., [Che.1, Secs. 7-3 and 7-4] that under state to input feedback (output to integrator-input feedback), controllability (observability

resp.) is maintained, so minimality can only be lost through loss of observability, (controllability resp.). Note also that (29) means that $\mathrm{rk}\begin{bmatrix} zI - A - BK \\ \hdashline C \end{bmatrix} < n$, whence the pair $(C, sI - A - BK)$ is not r.c.; hence it has a common nonunimodular right factor R s.t. $\det R(z) = 0$. As a result a pole-zero cancellation occurs in the transfer function $C(sI - A - BK)^{-1}B$. This transfer function has the same zeros as $H(s) = C(sI - A)^{-1}B$ iff minimality is conserved. A similar pole-zero cancellation occurs in $C(sI - A - KC)^{-1}B$ under (30).

32 Proof of Theorem 28. (a) Observe that $\begin{bmatrix} zI - A & B \\ - C & 0 \end{bmatrix}\begin{bmatrix} x_0 \\ -Kx_o \end{bmatrix} = \theta_{n+n_o}$, whence by (15) $z \in Z[H]$.

$\Rightarrow$: By assumption $z \in Z[H]$, whence by (15) $\exists$ a nonzero vector $(x_0^T, u_o^T)^T$ s.t.

33 $$\begin{bmatrix} zI - A & B \\ -C & 0 \end{bmatrix}\begin{bmatrix} x_0 \\ u_o \end{bmatrix} = \theta_{n+n_o}$$

Now $\mathrm{rk}(B) = n_i$ and (C, A) is observable. Hence both x_0 and u_o are nonzero; furthermore, B has a left inverse B^ℓ s.t. $B^\ell B = I_{n_i}$. Hence from the first equation of (33), $B^\ell(zI - A)x_0 + u_o = \theta$. So with $K := B^\ell(zI - A)$ we have from (33) $(A + BK)x_0 = zx_o$ and $Cx_o = \theta$.

(b) is left as an exercise. ∎

We conclude this section with a short consideration of poles and zeros at infinity.

35 Definition. [Kai.1, p.449]. Let $H \in \mathbb{R}(s)^{n_o \times n_i}$ and consider the map $s \in \mathbb{C} \mapsto \lambda = s^{-1} \in \mathbb{C}$; hence $s = \infty$ iff $\lambda = 0$. The poles and zeros of H at infinity are by definition the poles and zeros of the map $\lambda \in \mathbb{C} \mapsto H(\lambda^{-1})$ at $\lambda = 0$. ∎

Note that for observing the pole-zero structure at infinity we make a change of variables $s = \lambda^{-1}$ and investigate the SMM-form of the rational matrix $H(\lambda^{-1})$ at $\lambda = 0$.

36.R Exercise. Prove that if $D \in \mathbb{R}[s]^{n \times n}$ is a nonsingular column-reduced polynomial matrix, then D has no zero at ∞. (Hint: Consider $D(\lambda^{-1})$ and Theorem 6•••.)

37 Exercise. Prove: let $H \in \mathbb{R}(s)^{n_o \times n_i}$; then H is proper if and only if H has no pole at ∞. (Hint: Consider $H(\lambda^{-1})$ and a reasoning similar to that of Fact 3.)

38 Exercise. Prove: let $D \in \mathbb{R}[s]^{n\times n}$ be a nonsingular polynomial matrix; then D has no zeros at ∞ if and only if D^{-1} is proper. (Hint: A pole at ∞ of D^{-1} is a zero at ∞ of D•••.)

39 Fact [Pug.1, Thm 5]. Let $D \in \mathbb{R}[s]^{n\times n}$ be a nonsingular polynomial matrix; then D has no zeros at ∞ if and only if

40 $$\delta[\det D] = \delta_M[D],$$

where

$\delta_M[D]$ denotes the McMillan degree of the polynomial matrix D [Ros.1, pp.134-135, p. 117], and equals the maximal degree of any minor of any order of D.

41 Exercise. Show that

$$D(s) = \begin{bmatrix} 1 & s^3 & 0 \\ 0 & s & s \\ 0 & 0 & s^2 \end{bmatrix}$$

violates criterion (40). Hence D has a zero at ∞.

42 Remark. Criterion (40) is not very practical since it involves calculating the degrees of all minors of any order of the polynomial matrix D. Now in Remark 2.4.3.14 we mentioned that any nonsingular $D \in \mathbb{R}[s]^{n\times n}$ can be converted in a column-reduced matrix D_1 by e.c.o.'s •••. By Exercise 36 D_1 has no zeros at ∞, or equiv. D_1^{-1} is proper. This suggests a more practical criterion for having D^{-1} proper•••.

43.R Fact. Let $D \in \mathbb{R}[s]^{n\times n}$ be a nonsingular polynomial matrix and let $R \in \mathbb{R}[s]^{n\times n}$ be any unimodular matrix (obtained by e.c.o.'s) such that

44 $$DR = D_1,$$

where D_1 is c.r..

U.t.c.

45 D has no zeros at ∞ (equiv. D^{-1} is proper)

if and only if

46 $$\forall j \in \underline{n} \quad \partial_{cj}[R] \le \partial_{cj}[D_1].$$

47 Comment. "If R is cheap, then criterion (46) is cheap."

48 Exercise. Show that the matrix D of Exercise 41 does not satisfy criterion (46).

49 Exercise. Prove Fact 43.R (Hint: Use $RD_1^{-1} = D^{-1}$ and Theorem 2.4.3.25.R. ••••.)

50 Exercise. Get Fact 43.L (conversion to a row-reduced matrix, bounds on the row-degrees of the transforming matrix).

2.4.5. Dynamical Interpretation of Poles and Zeros

In this section we consider time-domain interpretations of transfer function-related definitions. The Laplace transform will be indicated by a superscript ^ or the symbol $\mathcal{L}[\cdots]$. The symbol $\mathcal{L}^{-1}[\cdots]$ is used for indicating the inverse Laplace transform.

1 Theorem [Poles]. Consider the transfer function matrix $\hat{H} \in \mathbb{R}(s)^{n_o \times n_i}$.
U.t.c.
If $p \in \mathcal{P}[\hat{H}]$, then $\exists$ an input $u(\cdot)$ which is a distribution with support at t=0 given by

2 $$u(t) = \sum_{k=0}^{\ell} u_k\, \delta^{(k)}(t), \text{ with } u_k \in \mathbb{C}^{n_i},\ \forall k = o \sim \ell$$

(whence

3 $$\hat{u}(s) = \sum_{k=0}^{\ell} u_k\, s^k \in \mathbb{C}[s]^{n_i}),$$

s.t. the output $\hat{y}(s) = \hat{H}(s)\hat{u}(s)$ satisfies

4 $$y(t) = \gamma \exp(pt) + \eta(t) \quad \forall t \geq 0$$

with γ a nonzero vector of $\mathbb{C}^{n_o}$, $\hat{\eta}(s) \in \mathbb{C}[s]^{n_o}$. ∎

5 Comment. The input $u(\cdot)$ "kicks" the system into the correct initial condition at $t = 0+$ such that, for $t > 0$, $y(t) = \gamma \exp(pt)$ (indeed, $\eta(\cdot)$ is a distribution with support at $t = 0$). "The input $u(\cdot)$, defined by (2), excites only the exponential mode p."

6 Proof of Theorem 1. Let (N_r, D_r) be a r.c.f. of $\hat{H}$. Since p is a pole of a coprime fraction, $\det D_r(p) = 0$ and $\exists$ a nonzero vector $\xi \in \mathbb{C}^{n_i}$ s.t. $D_r(p)\xi = \theta$; moreover, $[N_r(p)^T \vdots\, D_r(p)^T]^T$ has full column rank and so $\gamma := N_r(p)\xi \neq \theta$. Set now $\hat{u}(s) := D_r(s)\xi/(s - p)$ and observe that $\hat{u} \in \mathbb{C}[s]^{n_i}$. Moreover, $\hat{y}(s) = N_r(s)\xi/(s - p) = \gamma/(s - p) + (N_r(s) - N_r(p))\xi/(s - p)$, where $\hat{\eta}(s) := (N_r(s) - N_r(p))\xi/(s - p) \in \mathbb{C}[s]^{n_o}$. ∎

From now on we assume that the transfer function matrix $\hat{H} \in \mathbb{R}(s)^{n_o \times n_i}$ has a normal rank r s.t.

7 $$r = \mathrm{rk}\hat{H} = \min(n_o, n_i).$$

10 Theorem [Zeros, $n_i \leq n_o$]. Consider the transfer function matrix $\hat{H} \in \mathbb{R}(s)^{n_o \times n_i}$ where $n_i \leq n_o$ and (7) holds.

U.t.c.

If $z \in Z[\hat{H}]$, then $\exists$ a nonzero vector $\xi \in \mathbb{C}^{n_i}$ and $\hat{m}(s) \in \mathbb{C}[s]^{n_i}$ s.t. for the input given by

11 $$u(t) := \xi \exp(zt) + m(t) \quad \forall t \geq 0,$$

the output $\hat{y}(s) = \hat{H}(s)u(s)$ is given by

12 $$y(t) = n(t) \quad \forall t \geq 0$$

with
$$\hat{n}(s) \in \mathbb{C}[s]^{n_o}.$$

13 Comments. (a) Note that both $m(\cdot)$ and $n(\cdot)$ are distributions with support at $t = 0$. Hence $\forall t > 0$, the input $u(t) = \xi \exp(zt)$, and it produces

an output $y(t) = \theta$, $\forall t > 0$. We observe the <u>blocking property of the zero z</u>. For this reason, z is often called a <u>zero of transmission</u>.

(b) The distribution $m(\cdot)$ "kicks" the system such that <u>none</u> of the modes associated with the possible poles of $\hat{y}(s)$ is excited.

14 <u>Proof of Theorem 10</u>. Let (D_ℓ, N_ℓ) be a ℓ.c.f. of $\hat{H}$, whence $\exists\, U_\ell, V_\ell \in E(\mathbb{R}[s])$ s.t.

15 $$N_\ell U_\ell + V_\ell D_\ell = I_{n_o}.$$

Now by assumption $z \in Z[\hat{H}]$, hence, by Theorem 2.4.4.6, $rk[N_\ell(z)] < n_i$, whence

16 $\exists$ a nonzero vector $\xi \in \mathbb{C}^{n_i}$ s.t. $N_\ell(z)\xi = \theta_{n_o}$.

Therefore, also by (16)

17 $$\hat{\zeta}(s) := N_\ell(s)\xi/(s - z) = (N_\ell(s) - N_\ell(z))\xi/(s - z) \in \mathbb{C}[s]^{n_o}.$$

Define now

18 $$\hat{m}(s) := -U_\ell(s)\,\hat{\zeta}(s) \in \mathbb{C}[s]^{n_i},$$

19 $$\hat{u}(s) := \xi/(s - z) + \hat{m}(s).$$

Then by (18)-(19), (17), and (15)

$$\begin{aligned}\hat{y}(s) = \hat{H}(s)\hat{u}(s) &= D_\ell(s)^{-1}N_\ell(s)\,\{\xi/(s - z) - U_\ell(s)\,\hat{\zeta}(s)\}\\ &= D_\ell(s)^{-1}\,\{I - N_\ell(s)\,U_\ell(s)\}\,\hat{\zeta}(s)\\ &= D_\ell(s)^{-1}\,D_\ell(s)\,V_\ell(s)\,\hat{\zeta}(s)\\ &= V_\ell(s)\,\hat{\zeta}(s) =: \hat{n}(s) \in \mathbb{C}[s]^{n_o}.\end{aligned}$$ ∎

22 <u>Theorem</u> [Zeros, $n_o \le n_i$]. Consider the transfer function matrix $\hat{H} \in \mathbb{R}(s)^{n_o \times n_i}$ where $n_o \le n_i$ and (7) holds.

U.t.c.
If $z \in Z[\hat{H}]$ and $z \notin P[\hat{H}]$, then $\exists$ a nonzero vector $\eta \in \mathbb{C}^{n_o}$ such that for

every nonzero vector $\xi \in \mathbb{C}^{n_i}$ $\exists$ $\hat{m}_\xi(s) \in \mathbb{C}[s]^{n_i}$ s.t. the input

23 $$u(t) := \xi \exp(zt) + m_\xi(t) \qquad \forall t \geq 0$$

produces an output $\hat{y}(s) = \hat{H}(s)\hat{u}(s)$ satisfying

24 $$\eta^* y(t) = 0 \qquad \forall t \geq 0.$$

25 Comment. In contrast to Theorem 10, Theorem 22 allows any nonzero $\xi \in \mathbb{C}^{n_i}$ and asserts that the transmission of $\xi \exp(zt)$ has been blocked in the output direction defined by η: $\forall t > 0$ "y(t) can only belong to the hyperplane orthogonal to η, viz. $< \eta, y(t) > = 0$."

26 Proof of Theorem 22. Let (D_ℓ, N_ℓ) be a ℓ.c.f. of H. Since $z \in Z[H]$, by Theorem 2.4.4.6, $rk[N_\ell(z)] < n_o$, and

27 $\exists$ a nonzero vector $\gamma \in \mathbb{C}^{n_o}$ s.t. $\gamma^* N_\ell(z) = \theta^*_{n_i}$. Since (D_ℓ, N_ℓ) is ℓ.c., $rk[D_\ell(z) \;\; N_\ell(z)] = n_o$ and it follows from (27) that

28 $$\eta^* := \gamma^* D_\ell(z) \neq \theta^*_{n_o}.$$

For any nonzero $\xi \in \mathbb{C}^{n_i}$, let now

29 $$h_\xi(s) := \gamma^* D_\ell(z) D_\ell(s)^{-1} N_\ell(s) \xi \in \mathbb{C}(s).$$

Then by (27) and the fact that $z \notin P[H]$

30 $$h_\xi(z) = 0$$

s.t.

31 with (n,d) a coprime fraction of h_ξ, $n(z) = 0$, $d(z) \neq 0$.

Consequently, since n and d are polynomials,

32 $$\nu(s) := n(s)/(s - z) \in \mathbb{C}[s]$$

and

33 $$\exists\ \alpha \neq 0 \quad \text{s.t.} \quad -1 + \alpha d(z) = 0.$$

Hence by (33),

34 $$\pi(s) := (-1 + \alpha d(s))/(s - z) \in \mathbb{C}[s],$$

35 $$[1 + (s - z)\pi(s)]/d(s) \equiv \alpha \neq 0.$$

Using the ξ chosen above, define

36 $$\hat{m}_{\xi}(s) := \xi\pi(s) \in \mathbb{C}[s]^{n_i}$$

and

37 $$\hat{u}(s) := \xi/(s - z) + \hat{m}_{\xi}(s).$$

Then, using (28), (36)-(37), (29), (31), (32), and (35) it follows that

$$\begin{aligned} \eta^* \hat{y}(s) &= \gamma^* D_{\ell}(z)\, D_{\ell}(s)^{-1}\, N_{\ell}(s)\, \xi\, [(s - z)^{-1} + \pi(s)] \\ &= h_{\xi}(s)\, [(s - z)^{-1} + \pi(s)] \\ &= (n(s)/(s - z))\, \{[1 + (s - z)\pi(s)]/d(s)\} \\ &= \nu(s)\, \alpha \in \mathbb{C}[s]. \end{aligned}$$ ■

38 Exercise. Use the SMM-form equation $H = L\Psi_{\ell}^{-1}\mathcal{E}R$ to show that (30) follows from (29).

39 Comment. The dynamical interpretation of poles is fundamental to the understanding of stability questions (see Sec. 4.2). The dynamical interpretation of zeros is basic to the understanding of certain requirements imposed in the tracking problem (see Sec. 5.2). References for this section are [Des.1], [Cal.2].

2.5. Realization and Polynomial Matrix Fractions

This section is only a brief outline. For a complete treatment, see [Kai.1, Sec. 6.4, Sec. 6.5].

1.R Fact. Assume that we are given a PMD $[D_r, I, N_r, 0]$ corresponding to the equations

2 $$D_r(p)\xi(t) = u(t)$$

$$t \geq 0$$

3 $$y(t) = N_r(p)\xi(t)$$

and giving rise to a strictly proper transfer function satisfying

4 $$H = N_r D_r^{-1} \in \mathbb{R}_{p,o}(s)^{n_o \times n_i} \quad \text{s.t. } \underline{\operatorname{rk} H = n_i}$$

U.t.c. there exists a controllable realization $[A, B, C, 0]$ of H having the SSD

5 $$px(t) = Ax(t) + Bu(t)$$

$$t \geq 0,$$

6 $$y(t) = Cx(t)$$

where $x(t) \in \mathbb{R}^n$, with

7 $$n = \partial[\det D].$$ ∎

This is based on the following properties of (N_r, D_r):

(a) By Remark 2.4.3.14 and Theorem 2.3.3.18 we may assume without loss of generality that

8 $$D_r \in \mathbb{R}[p]^{n_i \times n_i} \text{ is column-reduced,}$$

s.t., if

9 $$k_j := \partial_{cj}[D_r] \qquad \forall j \in \underline{n}_i,$$

then

10 $$n := \partial[\det D] = \sum_{j=1}^{n_i} k_j.$$

(b) By (4) and Theorem 2.4.3.25.R,

11 $$0 \le \partial_{cj}[N_r] < \partial_{cj}[D_r] = k_j \quad \forall j \in \underline{n}_i .$$

(c) Because of (8)-(11) and Fact 2.4.3.21.R, $(N_r, D_r) \in \mathbb{R}[p]^{n_o \times n_i} \times \mathbb{R}[p]^{n_i \times n_i}$ can be represented by

12 $$D_r(p) = D_h S(p) + D_\ell \Psi(p),$$

13 $$N_r(p) = N_\ell \Psi(p),$$

where

14 $$S(p) := \operatorname{diag}[p^{k_j}]_{j=1}^{n_i} \in \mathbb{R}[p]^{n_i \times n_i},$$

15 $$\Psi(p) := \text{block diag}\left\{ \begin{bmatrix} p^{k_j-1} \\ \vdots \\ p \\ 1 \end{bmatrix}, k_j \times 1,\ j \in \underline{n}_i \right\} \in \mathbb{R}[p]^{n \times n_i},$$

16 $D_h \in \mathbb{R}^{n_i \times n_i}$ is the nonsingular highest column-degree coefficient matrix (2.4.3.15),

17 $D_\ell \in \mathbb{R}^{n_i \times n}$ is a coefficient matrix taking into account lower-degree terms of each entry of D_r,

18 $N_\ell \in \mathbb{R}^{n_o \times n}$ is a coefficient matrix taking into account the entries of N_r.

Procedure. Using (12)-(18), write equations (2)-(3) as

19 $$S(p)\xi(t) = -D_h^{-1} D_\ell \Psi(p) \xi(t) + D_h^{-1} u(t)$$

20 $$y(t) = N_\ell \Psi(p) \xi(t).$$

Hence a state $x(t) \in \mathbb{R}^n$ can be constructed as

21 $$x(t) := \Psi(p)\xi(t) = (p^{k_1-1}\xi_1(t), \cdots, p\xi_1(t), \xi_1(t); \cdots; p^{k_{n_i}-1}\xi_{n_i}(t), \cdots, p\xi_{n_i}(t), \xi_{n_i}(t))^T$$

using n_i integrator chains of length k_j driven by $p^{k_j}\xi_j(t)$, a first-order derivative of a component of the state $x(t)$ by (21). Note that by (19) and (21), $\forall j \in \underline{n}_i$, $p^{k_j}\xi_j(t)$ is an $\mathbb{R}$-linear combination of state and input components: this generates n_i equations of the state equation $px(t) = Ax(t) + Bu(t)$. This equation is then completed using $n - n_i$ integrator relations of the form $px_\ell(t) = x_{\ell-1}(t)$. The readout map $y(t) = Cx(t)$ follows from (20)-(21).

For example, let

$$D_r(p) = \begin{bmatrix} (p+1)^2 & p \\ 0 & p-1 \end{bmatrix}, \quad N_r(p) = \begin{bmatrix} p & 0 \\ 1 & 1 \end{bmatrix}.$$

Then $n_o = n_i = 2$, D_r is c.r., $k_1 = 2$, $k_2 = 1$, $n = k_1 + k_2 = 3$. (12)-(18) result in:

$$S(p) = \text{diag}[p^2, p],$$

$$\Psi(p) = \left[\begin{array}{c|c} p & 0 \\ 1 & 0 \\ \hline 0 & 1 \end{array}\right],$$

$$D_h = \begin{bmatrix} 1 & 1 \\ 0 & 1 \end{bmatrix},$$

$$D_\ell = \left[\begin{array}{cc|c} 2 & 1 & 0 \\ \hline 0 & 0 & -1 \end{array}\right], \quad N_\ell = \left[\begin{array}{cc|c} 1 & 0 & 0 \\ \hline 0 & 1 & 1 \end{array}\right].$$

Hence (19) reads

$$p^2\xi_1(t) = -2p\xi_1(t) - \xi_1(t) - \xi_2(t) + u_1(t) - u_2(t),$$

$$p\,\xi_2(t) = \xi_2(t) + u_2(t).$$

The state is constructed by (21) as

$$x(t) = (x_1(t),\ x_2(t);\ x_3(t))^T := (p\xi_1(t),\ \xi_1(t);\ \xi_2(t))^T$$

using $n_i = 2$ integrator chains of length $k_1 = 2$, $k_2 = 1$ resp. and driven by $p^2\xi_1(t) = px_1(t)$, $p\xi_2(t) = px_3(t)$ resp. Using two equations generated by (19) and (21) and one integrator relation, the state equation $px(t) = Ax(t) + Bu(t)$ is s.t.

$$A = \left[\begin{array}{cc|c} -2 & -1 & -1 \\ 1 & 0 & 0 \\ \hline 0 & 0 & 1 \end{array}\right], \quad B = \left[\begin{array}{cc} 1 & -1 \\ 0 & 0 \\ \hline 0 & 1 \end{array}\right].$$

From (20)-(21) we have a readout map $y(t) = Cx(t)$ with

$$C = \left[\begin{array}{cc|c} 1 & 0 & 0 \\ \hline 0 & 1 & 1 \end{array}\right].$$

Controllability. Note that the state equation $px(t) = Ax(t) + Bu(t)$ is obtained from the state equation $px(t) = A^o x(t) + B^o v(t)$ of the $n_i = 2$ integrator chains described by

$$A^o = \left[\begin{array}{cc|c} 0 & 0 & 0 \\ 1 & 0 & 0 \\ \hline 0 & 0 & 0 \end{array}\right], \quad B^o = \left[\begin{array}{cc} 1 & 0 \\ 0 & 0 \\ \hline 0 & 1 \end{array}\right], \quad (A^o, B^o) \text{ controllable, through}$$

the application of a feedback law given by

$$v(t) = D_h^{-1}(u(t) - D_\ell x(t)).$$

The latter represents an input-coordinate transformation followed by state feedback. As a result

$$A = A^o - B^o D_h^{-1} D_\ell \qquad B = B^o D_h^{-1},$$

as is easily checked. Since (A^o, B^o) is controllable it follows that (A, B) is controllable because the transformations preserve controllability. Notice finally that the above realization is observable and therefore minimal iff (N_r, D_r) is r.c. [Kai.1, Th. 6.5 -1].

23 Exercise. Develop a dual technique to obtain an observable realization of dimension $n = \partial[\det D_\ell]$ for the PMD $[D_\ell, N_\ell, I, 0]$ giving rise to a transfer

function $H = D_\ell^{-1} N_\ell \in \mathbb{R}_{p,o}(s)^{n_o \times n_i}$ s.t. rk $H = n_o$. Call this Fact 1.L [Kai.1, pp. 413-417].

24 Exercise. Study the relation between coprime fractions and minimal realizations in [Kai.1, pp. 439-440].

Chapter 3. Polynomial Matrix System Descriptions and Related Transfer Functions

3.1. Introduction

This chapter studies the time-domain properties of a polynomial matrix system description (PMD) and related algebraic properties of transfer functions. Section 2 studies the dynamics of a PMD (pseudo-state trajectory, response, ···) and the possibility of dynamical redundancy (reachability, observability, hidden modes, ···). Section 3 investigates the quality of the behavior at t = 0 of a PMD (well-formed PMDs, ···), and the exponential stability of a PMD (exponentially decreasing pseudo-state trajectory, ···). Section 4 describes properties of transfer functions generated by a PMD: right-left fractions and internally proper fractions (corresponding to well-formed PMDs).

3.2. Dynamics of a PMD; Redundancy

3.2.1. Dynamics of a PMD

Consider the PMD $[D, N_\ell, N_r, K]$ described by the equations

1 $$D(p)\xi(t) = N_\ell(p)u(t)$$

$$t \geq 0$$

2 $$y(t) = N_r(p)\xi(t) + K(p)u(t),$$

where (a) $D(p) \in \mathbb{R}[p]^{\nu\times\nu}$, $N_\ell(p) \in \mathbb{R}[p]^{\nu\times n_i}$, $N_r(p) \in \mathbb{R}[p]^{n_o\times\nu}$, $K(p) \in \mathbb{R}[p]^{n_o\times n_i}$;

(b) $D(\cdot)$ is nonsingular;

(c) $u(\cdot)$: $\mathbb{R}_+ \to \mathbb{R}^{n_i}$, $\xi(\cdot)$: $\mathbb{R}_+ \to \mathbb{R}^\nu$, $y(\cdot)$: $\mathbb{R}_+ \to \mathbb{R}^{n_o}$ are called the <u>input</u>, <u>pseudo-state</u>, and <u>output</u> of the PMD;

(d) $n := \partial[\det D(\cdot)] > 0$ is called the <u>order</u> of the PMD;

(d) $u(\cdot)$ is piecewise sufficiently differentiable (see the definition below). Note that the presence of the differentiation operator $p = d/dt$ in $N_\ell(p)$ and $K(p)$ requires the knowledge of the value of $u(\cdot)$ and a sufficient number of its derivatives at t = 0-.

3 A function $f : \mathbb{R}_+ \to \mathbb{R}^n$ is said to be piecewise sufficiently differentiable (p. suff. diff.) iff (a), except for a set $D \subset \mathbb{R}_+$, $f \in C^r$ (equiv. f is r times continuously differentiable, where r is large enough so that all differentiations are well defined); (b) D contains at most a finite number of points per unit interval; (c) $\forall \tau \in D$, $\forall k = 0 \sim r$, $f^{(k)}(\tau-)$ and $f^{(k)}(\tau+)$ are well defined and finite.

4 A typical example of a p. suff. diff. function (with r = 0) is $t \mapsto k(1(t) - 1(t-1))$ with $k \in \mathbb{R}^n$ and $1(\cdot)$ the unit step function.

5 We call state at t = 0 of the PMD described by (1)-(2), any state $x(0) \in \mathbb{R}^n$ of the differential equation $D(p)\xi(t) = \theta_\nu$ $t \geq 0$ (see Definition 2.3.5.16).

6 We call zero-input pseudo-state trajectory (z-i p-s trajectory) of the PMD any C^∞-solution $\xi(\cdot) : (0-, \infty) \to \mathbb{R}^\nu$ of the differential equation $D(p)\xi(t) = \theta_\nu$ $t \geq 0$. Hence by definition $\xi^{(j)}(0-) = \xi^{(j)}(0+)$ $\forall j \in \mathbb{N}$.

7 Among the states at t = 0 of the PMD we note the (normalized) state $x_\xi(0) \in \mathbb{R}^n$ of the PMD, which is defined as the normalized state $x_\xi(0) \in \mathbb{R}^n$ of the differential equation $D(p)\xi(t) = \theta_\nu$ $t \geq 0$ (see Definition 2.3.5.43).

8 Note that by, Algorithm 2.3.5.36, $x_\xi(0) \in \mathbb{R}^n$ is given in terms of $\xi(0)$ and some of its derivatives. Note that by Theorem 2.3.5.38 and Definition 6 the z-i p-s trajectory of the PMD has the form

9 $$\xi(t) = \psi(t)\, P x_\xi(0) \quad \forall t > 0-,$$

where $\Psi(\cdot)$ is any basis matrix (2.3.5.13) for the solution space X of the differential equation $D(p)\xi(t) = \theta_\nu$ $t \geq 0$, and $P \in \mathbb{R}^{n \times n}$ is a nonsingular matrix.

10 We call zero-input response (z-i response) of the PMD the response $y(\cdot) : \mathbb{R}_+ \to \mathbb{R}^{n_o}$ of the PMD when $u(t) \equiv \theta$ $\forall t > 0-$.

11 Note that by the substitution of (9) in (2) the z-i response of the PMD is given by

12 $$y(t) = N_r(p)\xi(t) = N_r(p)\psi(t)Px_\xi(0), \quad \forall t > 0-,$$

where differentiations are to be made in the usual sense because $\xi(\cdot)$ and all its derivatives are continuous on $t > 0-$.

15 We call <u>characteristic polynomial</u> of the PMD the polynomial det $D(\lambda)$, and <u>eigenvalue of the PMD</u> any $\lambda \in \mathbb{C}$ s.t. det $D(\lambda) = 0$.

16 By using Laplace Transform techniques it follows that any z-i p-s trajectory $\xi(\cdot)$ of the PMD is an sum of exponential polynomials in t of the form

17 $$\sum_{k=0}^{m} (a_k t^k)\exp(\lambda t) \quad \forall t > 0-,$$

where $a_k \in \mathbb{C}^\nu$ $\forall k = 0 \sim m$, $\lambda \in \mathbb{C}$ is an eigenvalue of the PMD, with the restriction that iff $\lambda \in \mathbb{C}\backslash\mathbb{R}$ then $\xi(\cdot)$ contains also the complex conjugate companion form

18 $$\sum_{k=0}^{m} (\bar{a}_k t^k)\exp(\bar{\lambda} t) \quad \forall t > 0-.$$

21 We call <u>zero-state pseudo-state trajectory</u> (z-s p-s trajectory) <u>of the PMD</u> the solution $\xi(\cdot)$ of the differential equation

1 $$D(p)\xi(t) = N_\ell(p)u(t) \quad t \geq 0$$

driven by a piecewise sufficiently differentiable input $u(\cdot) : \mathbb{R}_+ \to \mathbb{R}^{n_i}$ where both $\xi(\cdot)$ and $u(\cdot)$ and all their derivatives are zero at $t = 0-$ (more precisely $\xi^{(i)}(0-) = \theta_\nu$ and $u^{(j)}(0-) = \theta_{n_i}$ $\forall i = 0, 1, 2, \cdots$ and $\forall j = 0, 1, 2, \cdots$).

22 As a result the z-s p-s trajectory can be calculated as follows. Taking the Laplace transform on both sides, we obtain

23 $$\hat{\xi}(s) = D(s)^{-1}N_\ell(s)\hat{u}(s)$$

where by division Theorem 2.4.3.37.L

24 $$D(s)^{-1}N_\ell(s) = \sum_{\alpha=0}^{k} W_\alpha s^\alpha + \hat{W}(s),$$

where $\forall\alpha = 0 \sim k$ $W_\alpha \in \mathbb{R}^{\nu\times n_i}$, $\hat{W}(\cdot) \in \mathbb{R}_{p,o}(s)^{\nu\times n_i}$ (whence $t \mapsto W(t)$ $:= \mathcal{L}^{-1}[\hat{W}](t)$ is a function on $\mathbb{R}$ with $W(t) = 0$ $\forall t < 0$). As a consequence the z-s p-s trajectory of the PMD is given by

25 $$\xi(t) = \sum_{\alpha=0}^{k} W_\alpha p^\alpha u(t) + \int_{0-}^{t} W(t-\tau)u(\tau)\, d\tau \quad \forall t > 0-$$

where the required differentiations in (25) may have to be interpreted in the distribution sense ($p1(t) = \delta(t)$; $p^\alpha 1(t) = \delta^{(\alpha-1)}(t)$, $\cdots$). Note that, since $W(t) = 0$ for $t < 0$ and $u(t) = \theta$ for t in a sufficiently small interval $(-\varepsilon, 0)$, the convolution integral in (25) is zero on $(-\varepsilon, 0)$.

26 By the substitution of the z-s p-s trajectory (25) in the readout map (2) we obtain the zero-state response (z-s response) of the PMD to the input $u(\cdot)$ as

27 $$y(t) = N_r(p)\{\sum_{\alpha=0}^{k} W_\alpha p^\alpha u(t) + 1(t)\int_{0-}^{t} W(t-\tau)u(\tau)d\tau\} + K(p)u(t), \quad \forall t > 0-,$$

where the required differentiations may again have to be made in the distribution sense; the factor $1(t)$ is inserted in (27) to emphasize that, for $t \in (-\varepsilon, 0)$, the convolution integral equals zero.

Call

28 $$\hat{H}(s) = N_r(s)D(s)^{-1}N_\ell(s) + K(s)$$

the transfer function of the PMD $[D, N_\ell, N_r, K]$; then the Laplace transform of the z-s response reads

29 $$\hat{y}(s) = \hat{H}(s)\hat{u}(s).$$

Let us consolidate:

30 Remarks. (a) The z-i p-s trajectory $\xi(\cdot)$ and the z-i response $y(\cdot)$ of a PMD are C^∞-functions (see (9) and (12)). In particular $\xi(0-) = \xi(0) = \xi(0+)$, $y(0-) = y(0) = y(0+)$.

(b) Since the input $u(\cdot)$ is p. suff. diff. with discontinuity set D, the z-s p-s trajectory $\xi(\cdot)$ and z-s response $y(\cdot)$ of a PMD are, in general, distributions with $\mathbb{R}_+$ as support; in fact, these distributions are C^r-functions except, possibly, for points in D. Hence we may have $\xi(0-) \neq \xi(0+)$ and $y(0-) \neq y(0+)$; furthermore, since there may be a distribution at 0 in the RHS of (25) and (27), $\xi(\cdot)$ and $y(\cdot)$ may not be defined at $t = 0$. For example, with $\xi(\cdot)$ and all its derivatives zero at $t = 0-$, for

$$\begin{bmatrix} 1 & p^2 + p \\ 0 & p \end{bmatrix} \begin{bmatrix} \xi_1(t) \\ \xi_2(t) \end{bmatrix} = \begin{bmatrix} 0 \\ 1(t) \end{bmatrix},$$

the z-s p-s tractory is $\xi_1(t) = -1(t) - \delta(t)$; $\xi_2(t) = t1(t)$, for $t > 0-$.
(c) Concerning the z-s p-s trajectory $\xi(\cdot)$ and the z-s response $y(\cdot)$, note that the maps $u(\cdot) \mapsto \xi(\cdot)$ and $u(\cdot) \mapsto y(\cdot)$ are $\mathbb{R}$-linear.
(d) Suppose that $u(\cdot)$ is p. suff. diff., whence, for some $r \in \mathbb{N}$, $u(\cdot)$ is C^r $\forall t \in \mathbb{R}_+ \setminus D$ (where D is the set of discontinuity points of $u(\cdot)$). Then the z-s p-s trajectory $\xi(\cdot)$ is p. suff. diff. and $\xi(\cdot)$ is C^r on $\mathbb{R}_+ \setminus D$ if and only if $D(s)^{-1} N_\ell(s)$ is proper. (The same holds for the z-s response $y(\cdot)$ if and only if $\hat{H}(s)$ is proper.)

32 Using linearity and (9), (12), (25), and (27) the complete pseudo-state trajectory (p-s trajectory) and complete response of the PMD are given by

33 $$\xi(t) = \psi(t)Px_\xi(0-) + \sum_{\alpha=0}^{k} W_\alpha p^\alpha u(t) + \int_{0-}^{t} W(t-\tau)u(\tau)\, d\tau, \quad \forall t > 0-,$$

respectively:

34 $$y(t) = N_r(p)\left\{\psi(t)Px_\xi(0-) + \sum_{\alpha=0}^{k} W_\alpha p^\alpha u(t) + \int_{0-}^{t} W(t-\tau)u(\tau)\, d\tau\right\} + K(p)u(t), \quad \forall t > 0-.$$

In view of Remark 30, we wrote "$x_\xi(0-)$" in (33) and (34), rather than "$x_\xi(0)$," which was legitimately used in the z-i case.

35 Remark. Equations (33) and (34) imply the following important observation: For any p. suff. diff. $u(\cdot) : \mathbb{R}_+ \to \mathbb{R}^{n_i}$ satisfying $u^{(j)}(0-) = \theta_{n_i}$, $\forall j = 0, 1, 2, \cdots$ and for any state at 0-, $x_\xi(0-)$, the complete p-s trajectory and the complete response of the PMD are uniquely defined.

3.2.2. Reachability of PMDs

Consider the PMD $\mathcal{D} := [D, N_\ell, N_r, K]$ described by

1 $$D(p)\xi(t) = N_\ell(p)u(t)$$

$$t \geq 0$$

2 $$y(t) = N_r(p)\xi(t) + K(p)u(t)$$

with characteristics given in (3.2.1.1)-(3.2.1.2).

3 Definition. A z-i p-s trajectory $\xi(\cdot)$ of $\mathcal{D}$ is called reachable iff there is an input $u(\cdot)$ of the form $\sum_{\alpha=0}^{m} u_\alpha \delta^{(\alpha)}(t)$ (where $u_\alpha \in \mathbb{R}^{n_i}$ for $\alpha = 0 \sim m$) which produces a z-s p-s trajectory $\xi_1(\cdot)$ such that for $t > 0$, $\xi_1(t) = \xi(t)$; equivalently, iff $\exists$ an input $\hat{u}(s) \in \mathbb{R}[s]^{n_i}$ which produces a z-s p-s trajectory $\hat{\xi}_1(s) \in \mathbb{R}(s)^\nu$ such that

$$\hat{\xi}_1(s) = \hat{\xi}(s) + \hat{\eta}(s),$$

where $\hat{\eta}(s)$ is the polynomial part of $\hat{\xi}_1(s)$. (Consequently, $\hat{\xi}(s) \in \mathbb{R}_{p,o}(s)^\nu$ and $\mathcal{L}^{-1}[\hat{\eta}](t) = \theta_\nu \; \forall t > 0$.)

Analysis. Consider equation (1). Let $L(\cdot) \in \mathbb{R}[s]^{\nu\times\nu}$ be a g.c.ℓ.d. of (D, N_ℓ): equiv. $\exists$ polynomial matrices $\bar{D}$ and $\bar{N}_\ell$ s.t.

4 $$D = L\bar{D} \qquad N_\ell = L\bar{N}_\ell$$

and

5 $(\bar{D}, \bar{N}_\ell)$ is ℓ.c., equiv. $\text{rk}[\bar{D}(s) \,\vdots\, \bar{N}_\ell(s)] = \nu$, $\forall s \in \mathbb{C}$

(or equiv. for some $\bar{U}_\ell, \bar{V}_\ell \in \mathcal{E}(\mathbb{R}[s])$,

6 $\bar{N}_\ell\bar{U}_\ell + \bar{D}\bar{V}_\ell = I_\nu, \quad \forall s \in \mathbb{C}$)

10 Theorem. The z-i p-s trajectory $\xi(\cdot)$ of $\mathcal{D}$ is reachable

$\Leftrightarrow$

11 $$\bar{D}(p)\xi(t) = \theta_\nu \quad \forall t \geq 0.$$

12 Comment. (a) If (D, N_ℓ) is not ℓ.c., then the only z-i p-s trajectories that are reachable are those that satisfy (11). For example, if $\det L(p_0) = 0$ and $\det \bar{D}(p_0) \neq 0$, then by Theorem 10 the z-i p-s trajectory ξ_0 defined by

13 $$\xi_0(t) = ke^{p_0 t} \quad \text{where } k \in \mathbb{C}^\nu \text{ is s.t. } D(p_0)k = L(p_0)\,\bar{D}(p_0)\,k = \theta_\nu,$$

is not reachable.

(b) The set of reachable z-i p-s trajectories is a subspace of the linear space of z-i p-s trajectories.

14 Proof of Theorem 10

$\Rightarrow$: By assumption the z-i p-s trajectory $\xi(\cdot)$ is reachable, or equiv. for some $\hat{u} \in \mathbb{R}[s]^{n_i}$

15 $$D(s)\hat{\xi}_1(s) = N_\ell(s)\hat{u}(s),\ \hat{\xi}_1(s) = \hat{\xi}(s) + \hat{\eta}(s),\ \hat{\eta} \in \mathbb{R}[s]^\nu.$$

Now $\xi(\cdot)$ is a z-i p-s trajectory, equiv. $D(p)\xi(t) = \theta_\nu\ \forall t \geq 0$; hence for some polynomial vector $\pi(s) \in \mathbb{R}[s]^\nu$

16 $$D(s)\hat{\xi}(s) = \pi(s) \quad \text{and} \quad D(s)^{-1}\pi(s) \in \mathbb{R}_{p,o}(s)^\nu.$$

Using (15), (16), and (4), we obtain successively

$$D(s)[D(s)^{-1}\pi(s) + \hat{\eta}(s)] = N_\ell(s)\hat{u}(s)$$

$$\pi(s) = -D(s)\hat{\eta}(s) + N_\ell(s)\hat{u}(s) = L(s)\,[-\bar{D}(s)\hat{\eta}(s) + \bar{N}_\ell(s)\hat{u}(s)].$$

Hence defining

$$\bar{\pi}(s) := -\bar{D}(s)\hat{\eta}(s) + \bar{N}_\ell(s)\hat{u}(s),$$

where $\bar{\pi}(s)$ is a polynomial vector, we have

17 $$\pi(s) = L(s)\bar{\pi}(s),\ \bar{\pi}(s) \in \mathbb{R}[s]^\nu,$$

i.e., the initial conditions polynomial vector $\pi(\cdot)$ of (16) is divisible without remainder by $L(\cdot) \in \mathbb{R}[s]^{\nu\times\nu}$.

By (16), (17) and (4) we obtain now

18 $$\bar{D}(s)\hat{\xi}(s) = \bar{\pi}(s) \quad \text{and} \quad \hat{\xi}(s) = \bar{D}(s)^{-1}\bar{\pi}(s) \in \mathbb{R}_{p,o}(s)^{\nu}$$

Thus $\hat{\xi}(s)$ may be viewed as the response of $\bar{D}(s)^{-1}$, starting from the zero state at $t = 0-$ to the input $\bar{\pi}(s)$; note here that the Laplace transform takes into account values of $\xi(\cdot)$ on $\mathbb{R}_+$. Hence in the time domain, with $\bar{\pi}(s) := \sum_{\alpha=0}^{m} \pi_\alpha s^\alpha$, we have

19 $$\bar{D}(p)\xi(t) = \sum_{\alpha=0}^{m} \pi_\alpha \delta^{(\alpha)}(t) \qquad t \geq 0,$$

where at $t = 0-$ we have set $\xi^{(j)}(0-) = \theta_\nu \ \forall j = 0, 1, 2, \cdots$. Hence

20 $$\bar{D}(p)\xi(t) = \theta_\nu \ \forall t > 0,$$

i.e., $\xi(\cdot)$ is the solution of the differential equation $\bar{D}(p)\xi(t) = \theta_\nu$ $t \geq 0$ on $(0, \infty)$. Now the original function $t \mapsto \xi(t)$ is defined upon $(0-, \infty)$ and is a C^∞-extension of its restriction on $(0, \infty)$. So for this $\xi(\cdot)$ we must have

11 $$\bar{D}(p)\xi(t) = \theta_\nu \ \forall t \geq 0.$$ ∎

$\Leftarrow$: By assumption (11) holds; hence $\exists\ \bar{\pi}(s) \in \mathbb{R}[s]^\nu$ s.t.

21 $$\bar{D}(s)\hat{\xi}(s) = \bar{\pi}(s).$$

Now, from (6), we have

22 $$\bar{N}_\ell \bar{U}_\ell \bar{\pi} + \bar{D}\bar{V}_\ell \bar{\pi} = \bar{\pi}.$$

Also letting $\hat{u} := \bar{U}_\ell \bar{\pi}$, $\hat{\eta} := -\bar{V}_\ell \bar{\pi}$, and noting that (21) implies that

$$D(s)\hat{\xi}(s) = L(s)\bar{\pi}(s),$$

by the use of (4), we rewrite (22) after premultiplication by $L(\cdot)$ as

$$D(s)[\hat{\xi}(s) + \hat{\eta}(s)] = N_\ell(s)\hat{u}(s).$$

Hence by Definition 3 the input $\hat{u} \in \mathbb{R}[s]^{n_i}$ produces, for $t > 0$, the given z-i p-s trajectory $\xi(\cdot)$. (Note that the assumption, $\bar{D}(p)\xi(t) = \theta_\nu$, $\forall t \geq 0$, implies that $D(p)\xi(t) = L(p)\bar{D}(p)\xi(t) = \theta_\nu$, $\forall t \geq 0$.) ∎

From Theorem 10 we have now the following important theorem.

25 Theorem [Complete reachability of a PMD]. Consider the PMD $\mathcal{D}$ described by (1)-(2).
U.t.c.

26 Every z-i p-s trajectory $\xi(\cdot)$ of $\mathcal{D}$ is reachable

⇔

27 (D, N_ℓ) is ℓ.c.

28 Comment. Let L be a g.c.ℓ.d. of (D, N_ℓ) (see equations (4) and (5)). The meaning of Theorems 25 and 10 is that iff L is nonunimodular (equiv. (D, N_ℓ) is not ℓ.c.), then $\exists$ a z-i p-s trajectory of $\mathcal{D}$ which is not reachable or equiv. $\exists$ a solution $\xi(\cdot)$ of $D(p)\xi(t) = \theta_\nu$ $t \geq 0$ which is not a solution of $\bar{D}(p)\xi(t) = \theta_\nu$ $t \geq 0$: "some z-i p-s trajectory cannot be excited by means of an input": "there occurs a loss of nontrivial dynamics when (D, N_ℓ) has a nonunimodular common left factor and we must work in the zero-state input-excitation mode, $\hat{\xi}(s) = D(s)^{-1}N_\ell(s)\hat{u}(s)$." We obtain a dynamical interpretation of the common nonunimodular left factor L in the left fraction $D^{-1}N_\ell$ of the transfer function $\hat{H} = N_r D^{-1} N_\ell + K$ of the PMD.

29 Proof of Theorem 25. Because of (4)-(5), Theorem 10, and the definition of a z-i p-s trajectory, we are done if we can show that, with $L(\cdot)$ a g.c.ℓ.d. of (D, N_ℓ) and with X and $\bar{X}$ the solution spaces of

$$D(p)\xi(t) = \theta_\nu \quad t \geq 0 \quad \text{resp.} \quad \bar{D}(p)\bar{\xi}(t) = \theta_\nu \quad t \geq 0,$$

$L(\cdot)$ is unimodular

⇔

$$\bar{X} = X$$

⇒ : By (4) $D(p) = L(p)\bar{D}(p)$, where $L(\cdot) \in \mathbb{R}[p]^{\nu\times\nu}$; hence $\bar{X} \subset X$. Now by assumption $L(\cdot)$ is unimodular, such that $\bar{D}(p) = L(p)^{-1}D(p)$ with $L(\cdot)^{-1} \in \mathbb{R}[p]^{\nu\times\nu}$; hence $X \subset \bar{X}$.

⇐ : By (4) $D(p) = L(p)\bar{D}(p)$, where $L(\cdot) \in \mathbb{R}[p]^{\nu\times\nu}$. by Assumption $X = \bar{X}$, hence $\dim X = \dim \bar{X}$. Hence according to Theorem 2.3.5.2, $\partial[\det D] = \partial[\det \bar{D}]$.

Hence $\partial[\det L] = \partial[\det D] - \partial[\det \bar{D}] = 0$. Therefore, L is unimodular. ■

Complete reachability is finally related to the absence of input-decoupling zeros [Ros.1].

33 Definition. $z \in \mathbb{C}$ is called an input-decoupling zero (i-d zero) of the PMD $\mathcal{D}$ described by (1)-(2) iff given any g.c.ℓ.d. $L(\cdot) \in \mathbb{R}[p]^{\nu\times\nu}$ of (D, N_ℓ), $\det L(z) = 0$.

We have then

34 Corollary [Absence of i-d zeros]. Consider the PMD $\mathcal{D}$ described by (1)-(2). U.t.c.

26 Every z-i p-s trajectory $\xi(\cdot)$ of $\mathcal{D}$ is reachable ■

⇔

35 $\mathcal{D}$ has no i-d zeros.

36 Exercise. Prove Corollary 34.

37 Exercise. Show that $z \in \mathbb{C}$ is an i-d zero of the PMD $\mathcal{D}$ described by (1)-(2) iff $\mathrm{rk}[D(z) \mid N_\ell(z)] < \nu$.

38 Exercise. Show that the SSD [A, B, C, D] is completely controllable iff every z-i p-s trajectory of the PMD [pI - A, B, C, D] is reachable.

3.2.3. Observability of PMDs

We are given a PMD $\mathcal{D} := [D, N_\ell, N_r, K]$ described by

1 $$D(p)\xi(t) = N_\ell(p)u(t)$$

$$t \geq 0$$

2 $$y(t) = N_r(p)\xi(t) + K(p)u(t)$$

with characteristics (3.2.1.1)-(3.2.1.2), and we want to observe the p-s trajectory $\xi(\cdot)$ through the output $y(\cdot)$. Since the contributions of the input $u(\cdot)$ in (3.2.1.34) can be computed separately, we are lead to the following definition.

3 Definition. A z-i p-s trajectory $\xi(\cdot)$ of $\mathcal{D}$ is said to be unobservable iff the corresponding z-i response $y(\cdot)$ satisfies

4 $$y(t) = N_r(p)\xi(t) = \theta_{n_o} \quad \forall t > 0$$

or equivalently, $\xi(\cdot)$ is a solution of the differential equation

5 $$\begin{bmatrix} D(p) \\ \hline N_r(p) \end{bmatrix} \xi(t) = \theta_{\nu+n_o} \qquad \forall t > 0.$$

∎

Analysis. Consider equation (5). Let $R(\cdot) \in \mathbb{R}[s]^{\nu\times\nu}$ be a g.c.r.d. of (N_r, D): equiv. $\exists$ polynomial matrices $\bar{N}_r$ and $\bar{D}$ s.t.

6 $$D = \bar{D}R \qquad N_r = \bar{N}_r R$$

and

7 $$(\bar{N}_r, \bar{D}) \text{ is r.c., equiv. } \operatorname{rk}\begin{bmatrix} \bar{D}(s) \\ \hline \bar{N}_r(s) \end{bmatrix} = \nu, \quad \forall s \in \mathbb{C}$$

(or equiv., for some $\bar{U}_r, \bar{V}_r \in \mathcal{E}(\mathbb{R}[s])$,

8 $$\bar{U}_r\bar{N}_r + \bar{V}_r\bar{D}_r = I_\nu, \ \forall s \in \mathbb{C}).$$

10 Theorem. A z-i p-s trajectory $\xi(\cdot)$ is unobservable
$\Leftrightarrow$

11 $$R(p)\xi(t) = \theta_\nu \quad \forall t > 0.$$

12 Comment. (a) If (N_r, D) is not r.c., then any g.c.r.d. is not unimodular and according to Fact 2.3.5.50, the differential equation (11) has nontrivial solutions, whence there are nontrivial unobservable z-i p-s trajectories. For example, if $\det R(z_o) = 0$, then by Theorem 10, $t \mapsto k\exp(z_o t)$ with $R(z_o)k = \theta_\nu$ is an unobservable z-i p-s trajectory.

(b) The set of unobservable z-i p-s trajectories is a subspace of the linear space of z-i p-s trajectories.

13 Proof of Theorem 10

$\Leftarrow$: By assumption $R(p)\xi(t) = \theta_\nu \; \forall t > 0$, hence by (6) $\forall t > 0$

$$\begin{cases} \bar{D}(p)R(p)\xi(t) = D(p)\xi(t) = \theta_\nu \\ \bar{N}_r(p)R(p)\xi(t) = N_r(p)\xi(t) = \theta_{n_o} . \end{cases}$$

Hence $\xi(\cdot)$ is a solution of the differential equation (5), showing that (i) $\xi(\cdot)$ is a z-i p-s trajectory and (ii) $\xi(\cdot)$ is unobservable.

$\Rightarrow$: By assumption $\xi(\cdot)$ is unobservable, hence using (5)-(6) we obtain successively for $t > 0$

$$\theta_{\nu+n_o} = \begin{bmatrix} D(p) \\ \text{-----} \\ N_r(p) \end{bmatrix} \xi(t) = \begin{bmatrix} \bar{D}(p) \\ \text{-----} \\ \bar{N}_r(p) \end{bmatrix} R(p)\xi(t).$$

Using (8), i.e., premultiplying the last equation by $[\bar{V}_r \mid \bar{U}_r]$, we obtain (11). ∎

Theorem 10 leads immediately to the result below; its proof is suggested in Comment 12.

16 Theorem [Complete observability of a PMD]. Consider the PMD $\mathcal{D}$ described by (1)-(2).

U.t.c.

17 Every nontrivial z-i p-s trajectory $\xi(\cdot)$ of $\mathcal{D}$ is observable

$\Leftrightarrow$

18 (N_r, D) is r.c.

19 Comment. Let R be a g.c.r.d. of (N_r, D) (see equations (6)-(7)). The meaning of Theorems 16 and 10 is that iff R is nonunimodular (equiv. (N_r, D) is not r.c.), then $\exists$ a nontrivial unobservable z-i p-s trajectory $\xi(\cdot)$ or equiv. $\exists$ a nontrivial solution $\xi(\cdot)$ of $R(p)\xi(t) = \theta \; \forall t > 0$. "There occurs a loss of nontrivial dynamics when (N_r, D) has a nonunimodular common right factor and we observe only $y(\cdot)$." We obtain a dynamical interpretation for the cancellation of the common nonunimodular right factor R in the right fraction N_rD^{-1} of the transfer function $\hat{H} = N_rD^{-1}N_\ell + K$ of the PMD.

Complete observability is finally related to the absence of output-decoupling zeros [Ros.1].

22 Definition. $z \in \mathbb{C}$ is called an output-decoupling zero (o-d zero) of the PMD $\mathcal{D}$ described by (1)-(2) iff, given any g.c.r.d. $R(\cdot) \in \mathbb{R}[p]^{\nu\times\nu}$ of (N_r, D), $\det R(z) = 0$.

We have then

23 Corollary [Absence of o-d zeros]. Consider the PMD $\mathcal{D}$ described by (1)-(2). U.t.c.

17 Every nontrivial z-i p-s trajectory $\xi(\cdot)$ of $\mathcal{D}$ is observable

$\Leftrightarrow$

24 $\mathcal{D}$ has no o-d zeros.

25 Exercise. Prove Corollary 23.

26 Exercise. Show that $z \in \mathbb{C}$ is an o-d zero of the PMD $\mathcal{D}$ described by (1)-(2) iff $\operatorname{rk}\begin{bmatrix} D(z) \\ \hline N_r(z) \end{bmatrix} < \nu$.

27 Exercise. Show that the SSD [A, B, C, D] is completely observable iff every nontrivial z-i p-s trajectory of the PMD [pI - A, B, C, D] is observable.

3.2.4. Minimality, Hidden Modes, Poles, and Zeros

Consider a PMD $\mathcal{D} := [D, N_\ell, N_r, K]$ described by

1 $$D(p)\xi(t) = N_\ell(p)u(t)$$

$$t \geq 0$$

2 $$y(t) = N_r(p)\xi(t) + K(p)u(t)$$

with characteristics (3.2.1.1)-(3.2.1.2).

3 The PMD $\mathcal{D}$ is said to be minimal iff $\mathcal{D}$ has no decoupling zeros, or equiv. iff $\mathcal{D}$ has neither i-d zeros nor o-d zeros. We say also that $\mathcal{D}$ has no hidden modes.

4 Comment. According to Corollaries 3.2.2.34 and 3.2.3.23, Definition 3 means that every z-i p-s trajectory $\xi(\cdot)$ of $\mathcal{D}$ is reachable and every nontrivial z-i p-s trajectory $\xi(\cdot)$ of $\mathcal{D}$ is observable.

From Theorems 3.2.2.25 and 3.2.3.16 we have also

5 Theorem. The PMD $\mathcal{D}$ described by (1)-(2) is minimal iff (D, N_ℓ) is ℓ.c. and (N_r, D) is r.c. ∎

Minimality also gives the possibility of an easy characterization of poles and zeros of the transfer function $\hat{H}$ of a PMD $\mathcal{D}$. To avoid unnecessary redundancy in the PMD we shall assume that the polynomial matrices $\begin{bmatrix} N_\ell \\ \hline K \end{bmatrix}$ and $[N_r \mid K]$ have full normal column (resp. row) rank (avoiding "trivial inputs and outputs").

6 Theorem [Poles and zeros]. Consider a minimal PMD $\mathcal{D}$ described by (1)-(2) with transfer function

7 $$\hat{H} = N_r D^{-1} N_\ell + K,$$

where $n := \partial[\det D]$ and

8 $$\text{rk}\begin{bmatrix} N_\ell \\ \hline K \end{bmatrix} = n_i, \quad \text{rk}[N_r \mid K] = n_o$$

U.t.c.

(a)

9 $$p \in \mathcal{P}[\hat{H}] \Leftrightarrow \det D(p) = 0;$$

(b) if

10 $$P(s) = \left[\begin{array}{c|c} D(s) & N_\ell(s) \\ \hline -N_r(s) & K(s) \end{array}\right]$$

denotes the system matrix, then

11 $$z \in Z[\hat{H}] \Leftrightarrow \text{rk}[P(z)] < \nu + \min(n_o, n_i).$$

12 <u>Comments</u>. (a) It follows from criterion (9) and Definition 3.2.1.15 that for a minimal PMD $\mathcal{D}$, $\lambda \in \mathbb{C}$ is an eigenvalue of $\mathcal{D}$ iff λ is a pole of the transfer function $\hat{H}$ of $\mathcal{D}$. For an arbitrary PMD $\mathcal{D}$, an eigenvalue λ of $\mathcal{D}$ is either a decoupling zero or a pole of $\hat{H}$ (see below).

(b) If $\mathcal{D}$ were not minimal and we accept criterion (10), then, according to Excercise 3.2.2.37 and 3.2.3.26, decoupling zeros are zeros of $\hat{H}$. If $\mathcal{D}$ is minimal, this cannot happen: <u>the occurrence of a zero of the transfer function is not associated with unreachability or unobservability of the PMD $\mathcal{D}$</u>.

(c) Assumption (8) bears only on criterion (10).

15 <u>Proof of Theorem 6</u>. Since $\mathcal{D}$ is minimal it follows by Theorem 5 that $D^{-1}N_\ell \in \mathbb{R}(s)^{\nu\times n_i}$ is a ℓ.c.f. and $N_rD^{-1} \in \mathbb{R}[s]^{n_o\times\nu}$ is a r.c.f. Therefore, by Theorems 2.4.1.25L and 2.4.1.25R, we can associate with these fractions two generalized Bezout identities having similar properties as those encountered in the beginning of the proof of Theorem 2.4.4.12. The reasoning of this proof yields the desired results.

16 <u>Exercise</u>. Complete the proof of Theorem 6.

17 <u>Classification of the eigenvalues of a PMD</u>. Consider any PMD $\mathcal{D}$ described by (1)-(2). According to Definition 3.2.1.15, $\lambda \in \mathbb{C}$ is an eigenvalue of $\mathcal{D}$ iff $\det D(\lambda) = 0$. According to Definitions 3.2.2.33 and 3.2.3.22, decoupling zeros must be eigenvalues of $\mathcal{D}$. Note also that by successive g.c.d. extractions, e.g., [Kai.1, pp. 580-582], any PMD can be made minimal by canceling left nonunimodular common factors between D and N_ℓ and then right nonunimodular common factors between N_r and D. Obviously, this leaves the transfer function $\hat{H}$ unchanged and at the end of this process we get a new PMD $\bar{\mathcal{D}}$ which is minimal and has no decoupling zeros. The cancellation process has removed all decoupling zeros of $\mathcal{D}$. Hence by Theorem 6 and the reasoning above, we have for any PMD $\mathcal{D}$, using ordered lists in which elements are repeated according to their multiplicities in det D:

18 (eigenvalues of $\mathcal{D}$) = (decoupling zeros of $\mathcal{D}$) + (poles of $\hat{H}$), where $\hat{H}$ is the transfer function of $\mathcal{D}$.

For a complete treatment, see [Kai.1, pp.580-582].

3.3. Well-Formed and Exponentially Stable PMDs

3.3.1. Well-Formed PMDs

In this section we consider a PMD $\mathcal{D} := [D, N_\ell, N_r, K]$ described by (3.2.1.1)-(3.2.1.2) and want to avoid impulsive behavior at $t = 0$ in the p-s trajectory $\xi(\cdot)$ and response $y(\cdot)$, (a) for every possible value of $\xi(\cdot)$ and its derivatives at $t = 0-$ and (b) for every input $u(\cdot)$ whose Laplace transform $\hat{u}(s) \in \mathbb{R}_{p,o}(s)^{n_i}$ is strictly proper, or equiv. on $t > 0$ $u(\cdot)$ is a sum of exponential polynomials in t as in (3.2.1.17) (hence $u(\cdot) \in C^\infty$ on $t > 0$ ($u(\cdot)$ is infinitely differentiable)). By impulsive behavior at $t = 0$ we mean that the time function contains Dirac terms involving $\delta(\cdot)$, $\delta^{(1)}(\cdot)$, $\cdots$ giving rise to a polynomial vector in s when taking the Laplace transform.

Note that by allowing the value of $\xi(\cdot)$ and its derivatives to become arbitrary at $t = 0-$ we are not sure to "catch" a z-i p-s trajectory $\xi(\cdot)$ or a z-i p-s response $y(\cdot)$. To ensure this we introduced in Secs. 2.3.5 and 3.2.1 the notion of state $x(0)$. This will not be done here.

We begin our study by considering differential equations.

3 Preliminary Analysis. Consider the differential equation

4 $$D(p)\xi(t) = \theta,$$

where $p = d/dt$ is the differential operator, $D(\cdot) \in \mathbb{R}[p]^{\nu\times\nu}$ is nonsingular, and $\xi(\cdot)$ is time dependent on $\mathbb{R}_+$. Let

5 $$D(s) := D_0 s^k + D_1 s^{k-1} + \cdots + D_k,$$

where $D_i \in \mathbb{R}^{\nu\times\nu}$, $\forall i$. Solving (4) by the Laplace transform method for arbitrary initial values of $\xi(\cdot)$ and its derivatives at $t = 0-$, we get the equation

7 $$D(s)\hat{\xi}(s) = \pi(s),$$

where the initial conditions vector $\pi(s) \in \mathbb{R}[s]^\nu$ is given by

8 $$\pi(s) = [s^{k-1}I \mid s^{k-2}I \mid \cdots \mid I]\begin{bmatrix} D_0 & & \bigcirc \\ D_1 & D_0 & \\ \vdots & & \ddots \\ D_{k-1} & \cdots & D_0 \end{bmatrix}\begin{bmatrix} \xi(0-) \\ \xi^{(1)}(0-) \\ \vdots \\ \xi^{(k-1)}(0-) \end{bmatrix}$$

and the coefficient matrix on the RHS is called a Toeplitz matrix, e.g., [Kai.1], [Ver.1]. A key feature of the initial conditions vector $\pi(\cdot)$ is that it is a k-dimensional $\mathbb{R}^{\nu}$-linear combination of $\nu\times\nu$ polynomial matrices which are the quotients of successive divisions of D(s) by s^iI, where $i = 1, \cdots, k$. Indeed, we have $\forall i \in \underline{k}$

$$D(s) = s^iI(D_0s^{k-i} + D_1s^{k-i-1} + \cdots + D_{k-i}) + (D_{k-i+1}s^{i-1} + \cdots + D_k).$$

Hence (7)-(8) can be rewritten as

9 $$\hat{\xi}(s) = D(s)^{-1}\pi(s) = [s^{-1}I \mid s^{-2}I \mid \cdots \mid s^{-k}I]\begin{bmatrix} \xi(0-) \\ \xi^{(1)}(0-) \\ \vdots \\ \xi^{(k-1)}(0-) \end{bmatrix}$$

$$-D(s)^{-1}[s^{-1}I \mid s^{-2}I \mid \cdots \mid s^{-k}I]\begin{bmatrix} D_k & D_{k-1} & & D_1 \\ & D_k & & \vdots \\ & & \ddots & \vdots \\ \bigcirc & & & D_k \end{bmatrix}\begin{bmatrix} \xi(0-) \\ \xi^{(1)}(0-) \\ \vdots \\ \xi^{(k-1)}(0-) \end{bmatrix}.$$

Note that formula (9) gives the solution $\hat{\xi}(s)$ of the differential equation (4) for all initial values of $\xi(\cdot)$ and its derivatives at t = 0-. Formula (9) is the key to our main result.

10 Definition. We say that the differential equation (4) is well formed iff for all initial values of $\xi(\cdot)$ and its derivatives at t = 0- the solution $\xi(\cdot)$ does not contain a distribution at t = 0 of the form $\sum_{i=0}^{\ell} \xi_i\delta^{(i)}(t)$, where

$\xi_i \in \mathbb{R}^{\nu}$ $\forall i \in \underline{\ell}$, or equiv. in terms of the Laplace transform: $\hat{\xi}(\cdot)$ given by (9) is strictly proper for all values of $\xi(\cdot)$ and its derivatives at $t = 0-$. ¤

11 Comment. Definition 10 guarantees that for all initial values of $\xi(\cdot)$ and its derivatives at $t = 0-$, for $t > 0$ we shall be on a trajectory $\xi(\cdot)$ of the differential equation (4) without impulsive behavior at $t = 0$. Note that we do not guarantee that $\xi(\cdot)$ is continuous at $t = 0$, whence $\xi(0-)$ may be different from $\xi(0+)$ (see Exercises 22 and 23 below).

We have now the following result:

14 Theorem [Well-formed differential equations]. The differential equation (4) is well formed if and only if

15 $$D \in \mathbb{R}[s]^{\nu\times\nu} \text{ has no zeros at } \infty$$

or equiv.

16 $$D^{-1} \in \mathbb{R}_p(s)^{\nu\times\nu}.$$ ▪

17 Exercise. Let

$$D(p) = \begin{bmatrix} p & p^3 \\ 0 & p \end{bmatrix}.$$

Clearly, D^{-1} is not proper. Show that for $\xi_1(0-) = \xi_2(0-) = \xi_2^{(2)}(0-) = 0$, the solution of (4) for $t \geq 0$ is given by

$$\xi(t) = \begin{bmatrix} \delta(t)\xi_2^{(1)}(0-) \\ 0 \end{bmatrix},$$

i.e., $\xi(\cdot)$ has impulsive behavior at $t = 0$ $\forall \xi_2^{(1)}(0-) \neq 0$.

18 Exercise [Invariance of well-formedness under exponential weighting]. Consider differential equation (4). Let $a \in \mathbb{R}$ and $\eta(t) := \xi(t)e^{-at}$, whence in terms of the Laplace transform $\hat{\xi}(s+a) = \hat{\eta}(s)$. Let $\tilde{D}(s) := D(s+a)$ $=: \tilde{D}_0 s^k + \tilde{D}_1 s^{k-1} + \cdots + \tilde{D}_k$. Show that

(a) $D(p)\xi(t) = \theta_\nu \quad t \geq 0$

iff

19 $$\tilde{D}(p)\eta(t) = \theta_\nu \quad t \geq 0$$

(b) $\hat{\eta}(s)$ is given by the equation $\tilde{D}(s)\hat{\eta}(s) = \tilde{\pi}(s)$, where $\tilde{\pi}(s)$ is obtained from (8) by replacing D_j and $\xi^{(j)}(0-)$ by resp. $\tilde{D}_j$ and $\eta^{(j)}(0-)$, $\forall j = 0, 1, \cdots, k-1$.
(c) $\hat{\eta}(s)$ satisfies equation (9) by replacing $\pi(s)$, $D(s)$, D_j, $\xi^{(j)}(0-)$ by resp. $\tilde{\pi}(s)$, $\tilde{D}(s)$, $\tilde{D}_j$, $\eta^{(j)}(0-)$, with $j = 1, 2, \cdots, k$.
(d) The differential equation (4) is well formed iff the differential equation (19) is well formed.

20 Proof of Theorem 14. Notice that by Exercise 2.4.4.38, conditions (15) and (16) are equivalent. Now by Definition 10 we must show that (16) holds if and only if $\hat{\xi}(s)$, as given by (9), is strictly proper $\forall\xi^{(j)}(0-)$ for $j = 0, 1, 2, \cdots$.
If: Without loss of generality we may assume that $D(0) = D_k$ is nonsingular. Indeed, if this does not hold, then, since $D(\cdot)$ is nonsingular, $\exists\, a \in \mathbb{R}$ s.t. $D(a)$ is nonsingular. It suffices then to consider $\eta(t) := \xi(t)e^{-at}$, which satisfies (19) with $\tilde{D}(0) = D(a) = \tilde{D}_k$ nonsingular (see Exercise 18). Now, since by assumption, $\forall\xi^{(j)}(0-)$, $j = 0, 1, 2, \cdots$, $\hat{\xi}(s) \in \mathbb{R}_{p,o}(s)^\nu$, we may set $\xi^{(j)}(0-) = \theta\ \forall j = 1, 2, \cdots$, and obtain from (9)

$$D(s)^{-1}s^{-1}D_k\xi(0-) = s^{-1}\xi(0-) - \hat{\xi}(s) \quad \forall\xi(0-).$$

Since the RHS is strictly proper $\forall\xi(0-)$, it follows that the LHS is strictly proper $\forall\xi(0-)$ with D_k nonsingular. Hence $D(s)^{-1}s^{-1}$ is strictly proper, whence (16) must hold.

Only if: By assumption D^{-1} is proper, hence the RHS of (9) is strictly proper $\forall\xi^{(j)}(0-)$, $j = 0, 1, 2, \cdots$. Therefore, the LHS of (9) must have the same property. ∎

21 Remark. Observe that by Exercises 2.4.4.36.R and 2.4.4.36.L, D^{-1} is proper when the nonsingular polynomial matrix D is column- or row-reduced. Observe also that the same applies if D is in upper-triangular Hermite row form. Hence any differential equation (4) where D is such a matrix is well formed.

22 Exercise. Let the differential equation (4) be well formed. With $D(s)$

given by (5), show that

$$\xi(0+) = \xi(0-) - [D(s)^{-1}]_{s=\infty}[D_k\xi(0-) + D_{k-1}\xi^{(1)}(0-) + \cdots + D_1\xi^{(k-1)}(0-)].$$

(Hint: Use the initial value Theorem (e.g., [Kai.1, p. 12]), viz., $\xi(0+) = \lim_{s\to\infty} s\,\hat{\xi}(s)$, and (9)).

23 Exercise. Let the differential equation (4) be well formed. Show that, for all initial values at $t = 0-$ of $\xi(\cdot)$ and all its derivatives, $\xi(\cdot)$ is continuous at $t = 0$ (equiv. $\xi(0-) = \xi(0+)$) if and only if D^{-1} is strictly proper. (Hint: Use the initial value Theorem, and argue as in the proof of Theorem 14.)

24 Exercise. Show that when differential equation (4) is s.t. D^{-1} is strictly proper, then (a) $\xi(0) := \xi(0+) = \xi(0-)$ is a partial state of (4), or equiv., with $n := \partial[\det D]$, there exists a state $x(0) \in \mathbb{R}^n$ (see Definition 2.3.5.16) such that $x(0) = [\xi(0)^T; x_2(0)^T]^T$, and (b) $n \geq \nu$. (Hint: Use $\xi(t) = \Psi(t)x(0)$, where $\Psi(0) : x(0) \to \xi(0)$ is now surjective; hence modulo a coordinate transformation $\bar{x}(0) = Px(0)$, $\det P \neq 0$, we get $\Psi(0)P^{-1} = [I_\nu \;\; 0]\cdots$).

Consider now a PMD $\mathcal{D} := [D, N_\ell, N_r, K]$ described by (3.2.1.1)-(3.2.1.2).

25 Definition. We say that the PMD $\mathcal{D}$ is zero-input well formed (z-i well formed) iff for all initial values of $\xi(\cdot)$ and its derivatives at $t = 0-$, the z-i p-s trajectory $\xi(\cdot)$ and the z-i response $y(\cdot)$ do not contain a distribution at $t = 0$ of the form $\sum_{i=0}^{\ell} \xi_i\delta^{(i)}(t)$, resp. $\sum_{j=0}^{m} y_j\delta^{(j)}(t)$, where, $\forall i$ and j, $\xi_i \in \mathbb{R}^\nu$ resp. $y_j \in \mathbb{R}^{n_o}$.

26 Analysis. The z-i p-s trajectory $\xi(\cdot)$ and the z-i response $y(\cdot)$ of $\mathcal{D}$ satisfy

27 $$D(p)\xi(t) = \theta_\nu \qquad \forall t \geq 0,$$

28 $$y(t) = N_r(p)\xi(t)$$

where these equations can be rewritten as a differential equation

28 $$\begin{bmatrix} I & -N_r(p) \\ 0 & D(p) \end{bmatrix} \begin{bmatrix} y(t) \\ \xi(t) \end{bmatrix} = \begin{bmatrix} \theta_{n_o} \\ \theta_\nu \end{bmatrix} \quad \forall t \geq 0.$$

Since obviously the initial values of $y(\cdot)$ and its derivatives at $t = 0-$ are not needed, it follows immediately that the PMD $\mathcal{D}$ is z-i well formed iff the differential equation (28) is well formed. Hence by Theorem 14 and Fact 2.4.4.39, we have

29 Theorem [z-i well formed PMDs]. The PMD $\mathcal{D}$ described by (3.2.1.1) and (3.2.1.2) is z-i well formed if and only if

30 D^{-1} and N_rD^{-1} are proper rational matrices or equivalently

31 $$\partial[\det D] = \delta_M\left(\begin{bmatrix} N_r \\ -- \\ D \end{bmatrix}\right),$$

where as in Fact 2.4.4.39, δ_M denotes the McMillan degree of the polynomial matrix between the braces and equals the highest degree of any minor of any order. ■

From Exercise 2.4.4.36.R and Theorem 2.4.3.25.R we have

32 Corollary. Consider a PMD $\mathcal{D}$ described by (3.2.1.1) and (3.2.1.2). Then $\mathcal{D}$ is z-i well formed if
(a) D is column reduced,
(b) $\forall j \in \underline{\nu}$, $\partial_{cj}[N_r] \leq \partial_{cj}[D]$. ■

A sharper (and more practical) result is, however, still available.

33 Corollary. Consider a PMD $\mathcal{D}$ described by (3.2.1.1) and (3.2.1.2). Let, as in Remark 2.4.3.14, $R \in \mathbb{R}[s]^{\nu\times\nu}$ be any unimodular matrix, obtained by e.c.o.'s on $\begin{bmatrix} D \\ \overline{N_r} \end{bmatrix}$, s.t.

$$DR = \bar{D} \quad \text{and} \quad N_rR = \bar{N}_r,$$

where $\bar{D}$ is c.r.

U.t.c.

$\mathcal{D}$ is z-i well formed
if and only

(a) $\forall j \in \underline{\nu} \quad \partial_{cj}[R] \le \partial_{cj}[\bar{D}]$,

(b) $\forall j \in \underline{\nu} \quad \partial_{cj}[\bar{N}_r] \le \partial_{cj}[\bar{D}]$. ∎

34 Exercise. Prove Corollary 33. (Hint: Use (30), Fact 2.4.4.43.R, and Theorem 2.4.3.25.R.)

35 Definition. We say that PMD $\mathcal{D} = [D, N_\ell, N_r, K]$, described by (3.2.1.1) and (3.2.1.2), is zero-state well formed (z-s well formed) iff every input $u(\cdot)$ with Laplace transform $\hat{u} \in \mathbb{R}_{p,o}(s)^{n_i}$ produces a z-s p-s trajectory $\xi(\cdot)$ and a z-s response $y(\cdot)$ such that their Laplace transforms satisfy $\hat{\xi} \in \mathbb{R}_{p,o}(s)^{\nu}$ and $\hat{y} \in \mathbb{R}_{p,o}(s)^{n_o}$. (Note that the class of time functions, whose Laplace transform is a strictly proper rational vector, is precisely the class of C^∞-functions on $t > 0$ that are sums of exponential polynomials in t as in (3.2.1.17)).

36 Analysis. Note that for a PMD $\mathcal{D}$, described by (3.2.1.1) and (3.2.1.2), the Laplace transforms of the z-s p-s trajectory $\xi(\cdot)$ and z-s response $y(\cdot)$ are given by

37 $$\hat{\xi} = D^{-1}N_\ell\hat{u}$$

38 $$\hat{y} = \hat{H}\hat{u}$$

where

39 $$\hat{H} = N_rD^{-1}N_\ell + K$$

is the transfer function of the PMD $\mathcal{D}$.

Note also the following dynamical interpretation of a pole at ∞ (see Definition 2.4.4.35).

40 Theorem [Dynamical interpretation of a pole at ∞]. Let $\hat{H} \in \mathbb{R}(s)^{n_o \times n_i}$ be a given transfer function.

U.t.c.

$\hat{H}$ has a pole at ∞

if and only if

there exists an input $u(\cdot)$ whose Laplace transform $\hat{u} \in \mathbb{R}_{p,o}(s)^{n_i}$ which produces an output $\hat{y} = \hat{H}\hat{u} \in \mathbb{R}(s)^{n_o}$ s.t. $\hat{y}$ has a polynomial part; more precisely $\hat{y}$ is not a strictly proper rational vector (equiv. for that input,

$y(\cdot)$ will have a nonzero distribution at $t = 0$).

41 Exercise. Prove Theroem 40. (Hint: Use Exercise 2.4.4.37.)

By (37), (38), Exercise 2.4.4.37, and Theorem 40, we obtain the following result.

42 Theorem [z-s well-formed PMDs]. A PMD $\mathcal{D} = [D, N_r, N_\ell, K]$ described by (3.2.1.1) and (3.2.1.2) is z-s well formed if and only if

43 $$D^{-1}N_\ell \in \mathbb{R}_p(s)^{\nu\times n_i}$$

and

44 $$\hat{H} := N_r D^{-1} N_\ell + K \in \mathbb{R}_p(s)^{n_o\times n_i},$$

where H is the transfer function of the PMD. ∎

We then have, finally:

47 Definition. We say that the PMD $\mathcal{D} = [D, N_\ell, N_r, K]$, described by (3.2.1.1) and (3.2.1.2), is well formed iff for every initial value of $\xi(\cdot)$ and its derivatives at $t = 0-$, and for every input $u(\cdot)$ such that (a) $u^{(j)}(0-) = \theta_{n_i}$ $\forall j = 0, 1, 2, \cdots$, and (b) $\hat{u} \in \mathbb{R}_{p,o}(s)^{n_i}$, we have that the p-s state trajectory $\xi(\cdot)$ and response $y(\cdot)$ of $\mathcal{D}$ satisfy $\hat{\xi} \in \mathbb{R}_{p,o}(s)^{\nu}$ and $\hat{y} \in \mathbb{R}_{p,o}(s)^{n_o}$ (equiv. they must not contain a distribution at $t = 0$).

48 Analysis. Because of linearity of the responses and Definitions 25 and 35: the PMD $\mathcal{D}$ is well formed iff (a) it is z-i well formed and (b) it is z-s well formed. So by Theorems 29 and 42, $\mathcal{D}$ is well formed iff the rational matrices D^{-1}, $N_r D^{-1}$, $D^{-1}N_\ell$, $\hat{H} = N_r D^{-1} N_\ell + K$ are proper. These four matrices are block entries of an extended transfer function $\hat{H}_e$ [Kai.1, pp. 569-570], given by

49 $$\hat{H}_e := \left[\begin{array}{c|c|c} I_{n_o} & N_r D^{-1} & -\hat{H} \\ \hline 0 & D^{-1} & -D^{-1}N_\ell \\ \hline 0 & 0 & I_{n_i} \end{array}\right].$$

Note also that $\hat{H}_e^{-1}$ is the polynomial matrix

50 $$\hat{H}_e^{-1} = \begin{bmatrix} I_{n_o} & -N_r & K \\ 0 & D & N_\ell \\ 0 & 0 & I_{n_i} \end{bmatrix}.$$

Hence from (49)-(50) and Exercise 2.4.4.38, the PMD $\mathcal{D}$ is well formed iff $\hat{H}_e^{-1}$ has no zeros at ∞. Note now that the system-matrix P of $\mathcal{D}$ [Ros.1] is given by

51 $$P = \begin{bmatrix} D & N_\ell \\ -N_r & K \end{bmatrix};$$

i.e., modulo a block-row permutation, $\hat{H}_e^{-1}$ is an extended system matrix obtained by bordering the system matrix appropriately by 0 or I. Using Fact 2.4.4.39, we have then that $\mathcal{D}$ is well formed iff $\partial[\det D] = \delta_M[P]$ = the highest degree of any minor of any order of P (exercise).

Hence we have obtained

52 Theorem [Well-formed PMDs]. Consider a PMD $\mathcal{D} := [D, N_r, N_\ell, K]$ described by (3.2.1.1) and (3.2.1.2), with transfer function matrix $\hat{H} \in \mathbb{R}(s)^{n_o \times n_i}$ and system matrix $P \in E(\mathbb{R}[s])$ given by

53 $$\hat{H} = N_r D^{-1} N_\ell + K$$

and

51 $$P = \begin{bmatrix} D & N_\ell \\ -N_r & K \end{bmatrix}.$$

U.t.c.

54 $\mathcal{D}$ is well formed

⇔

55 the rational matrices D^{-1}, $N_r D^{-1}$, $D^{-1} N_\ell$, $\hat{H}$ are proper

⇔

56 $$\partial[\det D] = \delta_M[P],$$

where δ_M denotes the McMillan degree of the polynomial matrix P and is the highest degree of any minor of any order of P. ∎

57 Exercise. Consider the scalar case, i.e., a PMD $\mathcal{D} = [d, n_\ell, n_r, k]$, where all elements are polynomials. Check that (56) ⇔ (55).

58 Comment. Condition (56) is a degree-limiting condition in order that (55) may hold: "Given a D s.t. D^{-1} is proper, there is, roughly speaking, a bound on the degrees of elements of N_r, N_ℓ, and K."

It is difficult to check condition (56). However a sufficient condition for checking condition (55) is readily available. For this we need the notion of a row-column-reduced polynomial matrix.

60 Definition. Let $D \in \mathbb{R}[s]^{\nu\times\nu}$ be a nonsingular polynomial matrix; then D is said to be row-column-reduced (r.c.r.) iff there exist integers $r_i \geq 0$, $i \in \underline{\nu}$, and $k_j \geq 0$, $j \in \underline{\nu}$, s.t.

61 $$\lim_{s\to\infty} \text{diag}[s^{-r_i}]_{i=1}^{\nu}\, D(s)\, \text{diag}[s^{-k_j}]_{j=1}^{\nu} = D_h \in \mathbb{R}^{\nu\times\nu} \text{ with } \det D_h \neq 0.$$

The integers r_i and k_j are called row powers, resp. column powers, and D_h is called the highest degree coefficient matrix.

62 Comment. Note that condition (61) is equivalent to either one of the two following factorizations of $D \in \mathbb{R}[s]^{\nu\times\nu}$:

63 $$D = \text{diag}[s^{r_i}]_{i=1}^{\nu}\, D_-\, \text{diag}[s^{k_j}]_{j=1}^{\nu}$$

or

64 $$D = D_\ell\, D_-\, D_r,$$

where $D_- \in \mathbb{R}(s)^{\nu\times\nu}$ is biproper (equiv. D_- and D_-^{-1} are proper), and $D_\ell \in \mathbb{R}[s]^{\nu\times\nu}$ and $D_r \in \mathbb{R}[s]^{\nu\times\nu}$ are row-reduced with row-degrees r_i, $i \in \underline{\nu}$, resp. column-reduced with column-degrees k_j, $j \in \underline{\nu}$. Moreover, according to Fact 2.4.3.21, a r.r. matrix $D \in \mathbb{R}[s]^{\nu\times\nu}$ is r.c.r. with column powers zero and similarly a c.r. matrix $D \in \mathbb{R}[s]^{\nu\times\nu}$ is r.c.r. with row powers zero. ∎

65 Fact. Let $D \in \mathbb{R}[s]^{\nu\times\nu}$ be a nonsingular polynomial matrix with entries $d_{ij} \in \mathbb{R}[s]$, where $i, j \in \underline{\nu}$.

U.t.c.

66 D is r.c.r. with row powers $r_i \geq 0$, $i \in \underline{\nu}$, column powers $k_j \geq 0$, $j \in \underline{\nu}$, and D_h in (61) diagonal

if and only if

67 $$\forall i \in \underline{\nu} \quad \partial[d_{ii}] = r_i + k_i,$$
$$\forall i,j \in \underline{\nu}, \text{ with } i \neq j \quad \partial[d_{ij}] < r_i + k_j. \qquad ✖$$

Proof. Use condition (63) with $D_h = D_-(\infty)$ diagonal. ✖

A nonsingular matrix $D \in \mathbb{R}[s]^{\nu\times\nu}$ can be made r.c.r. by e.o.'s over $\mathbb{R}[s]$···.

68 Remark. Any nonsingular matrix $D \in \mathbb{R}[s]^{\nu\times\nu}$ can be made c.r. by the method of Remark 2.4.3.14 by e.c.o.'s. Moreover, column permutations and premultiplication by a nonsingular constant matrix will further make D c.r. with column degrees $c_1 \geq c_2 \geq \cdots \geq c_\nu \geq 0$ and with highest column degree coefficient matrix $\bar{D}_h = I$. For example, $D \in \mathbb{R}[s]^{\nu\times\nu}$ has been converted into

69 $$\bar{D}(s) = \begin{bmatrix} s^5 & s & s \\ s^4 & s^3 & 1 \\ s^2 & s^2 & s^2 \end{bmatrix},$$

where $\nu = 3$, $c_1 = 5$, $c_2 = 3$, $c_3 = 2$, and $\bar{D}_h = I$. It is a fact that any c.r. matrix $\bar{D} \in \mathbb{R}[s]^{\nu\times\nu}$ with column degrees $c_1 \geq c_2 \geq \cdots \geq c_\nu \geq 0$ and $\bar{D}_h = I$ can, by e.c.o.'s, be converted into a r.c.r. matrix $\tilde{D}$ with $\tilde{D}_h = I$ and with row and column powers given by any pair of ν-tuples of integers $r_i \geq 0$, $i \in \underline{\nu}$, resp. $k_j \geq 0$, $j \in \underline{\nu}$ s.t.

70 $$r_1 \geq r_2 \geq \cdots \geq r_\nu \geq 0,\ k_1 \geq k_2 \geq \cdots \geq k_\nu \geq 0,\ r_i + k_i = c_i,\ \forall i \in \underline{\nu}.$$

The method [Ros.2, p. 841, line 19 from below] performs e.c.o.'s on columns $j = 1, 2, \cdots, \nu-1$ of $\bar{D}$; these e.c.o.'s are conditioned by the replacement of an entry (i,j), $i > j$, by its remainder after division by entry (i, i). This is done in such a way that the degrees of entries (i, j), $\forall i > j$, are lowered below those of entries (i, i), while the degrees of entry (j, j) and entries (i, j), $\forall i < j$, are maintained constant resp. below the degree of entry (j, j); moreover, each diagonal entry remains monic. Hence using (70) there results a matrix $\tilde{D}$ s.t.

$$\forall i \quad \partial[\tilde{d}_{ii}] = c_i = r_i + k_i,$$
$$\forall i > j \quad \partial[\tilde{d}_{ij}] < c_i = r_i + k_i \le r_i + k_j,$$
$$\forall i < j \quad \partial[\tilde{d}_{ij}] < c_j = r_j + k_j \le r_i + k_j.$$

As a result, condition (67) is satisfied and by Fact 65 we have that $\tilde{D}$ is r.c.r. with row and column powers r_i, resp. k_j, given by (70); moreover $\tilde{D}_h = I$.

To illustrate this method, note that the matrix $\bar{D}$, given in (69), is transformed by the e.c.o.'s $\gamma_1 \leftarrow \gamma_1 - s\gamma_2$, $\gamma_1 \leftarrow \gamma_1 + (s-1)\gamma_3$, $\gamma_2 \leftarrow \gamma_2 - \gamma_3$ into

$$\tilde{D}(s) = \begin{bmatrix} s^5 - s & 0 & s \\ s - 1 & s^3 - 1 & 1 \\ 0 & 0 & s^2 \end{bmatrix}$$

where $\tilde{D}$ is now r.c.r. with $\tilde{D}_h = I$ and row and column powers r_i resp. k_j conditioned by

$$r_1 \ge r_2 \ge r_3 \ge 0,\ k_1 \ge k_2 \ge k_3 \ge 0,\ r_1 + k_1 = 5,\ r_2 + k_2 = 3,\ r_3 + k_3 = 2.$$

From the arguments above it is also clear that, by using e.r.o.'s, any r.r. matrix $\bar{D}$ with decreasing row-degrees and $\bar{D}_h = I$ can be converted into a r.c.r. matrix with $\tilde{D}_h = I$ and row and column powers conditioned similarly as in (70).

71 <u>Remark</u>. Criterion (64) suggests another method for making a nonsingular matrix $D \in \mathbb{R}[s]^{\nu\times\nu}$ r.c.r. by e.o.'s over $\mathbb{R}[s]$.

1. Factorize D as

72 $$D = \bar{D}_\ell\, D_-\, \bar{D}_r,$$

where $\bar{D}_\ell$ and $\bar{D}_r \in \mathbb{R}[s]^{\nu\times\nu}$ and $D_- \in \mathbb{R}(s)^{\nu\times\nu}$ is <u>biproper</u>.

2. Get $\bar{D}_\ell$ r.r. by e.r.o.'s and $\bar{D}_r$ c.r. by e.c.o.'s.

Every nonsingular matrix $D \in \mathbb{R}[s]^{\nu\times\nu}$ has many factorizations (72) since it can be factored as a product of polynomial matrices in many ways (then $D_- = I$ in (72)).

Factorization (72) is useful in the study of feedback systems: the inverse return difference $[I + G_1G_2]^{-1}$ with $G_1 \in \mathbb{R}_p(s)^{m\times n}$ given by a ℓ.f. (D_1, N_1) and $G_2 \in \mathbb{R}_{p,o}(s)^{n\times m}$ given by a r.f. (N_2, D_2) satisfies

73 $$(I + G_1G_2)^{-1} = D_2\, D^{-1}\, D_1,$$

where

74 $$D = D_1D_2 + N_1N_2 = D_1(I + G_1G_2)D_2,$$

with $G_1G_2 \in \mathbb{R}_{p,o}(s)^{m\times m}$. Hence D is of the form (72) and by criterion (64), D is r.c.r. if D_1 is r.r. and D_2 is c.r.

75 Exercise. Show that every nonsingular diagonal matrix $D \in \mathbb{R}[s]^{\nu\times\nu}$ is r.c.r. ∎

The following property is important in that it relates $\partial[\det D]$ and the sum of row and column powers.

76 Fact. A nonsingular matrix $D \in \mathbb{R}[s]^{\nu\times\nu}$ is r.c.r. with row powers $r_i \geq 0$, $i \in \underline{\nu}$, and column powers $k_j \geq 0$, $j \in \underline{\nu}$, if and only if

$$\partial[\det D] = \sum_{i=1}^{\nu} (r_i + k_i).$$ ∎

77 Exercise. Prove Fact 76. (Hint: (a) Argue on the $r_i \geq 0$, $i \in \underline{\nu}$ and the $k_j \geq 0$, $j \in \underline{\nu}$ s.t. $\lim_{s\to\infty} \operatorname{diag}[s^{-r_i}]\, D(s)\, \operatorname{diag}[s^{-k_j}] = D_h \in \mathbb{R}^{\nu\times\nu}$, (b) use $\lim \det \cdots = \det \lim \cdots$.)

We return now to the study of well-formed PMDs.

78 Fact. A nonsingular matrix $D \in \mathbb{R}[s]^{\nu\times\nu}$ which is r.c.r. is such that D^{-1} is proper, or equiv. D has no zeros at ∞.

79 Exercise. Prove Fact 78. (Hint: $D(s)^{-1} = \operatorname{diag}[s^{-k_j}]\cdot$ $\left(\operatorname{diag}[s^{-r_i}]\, D(s)\, \operatorname{diag}[s^{-k_j}]\right)^{-1} \operatorname{diag}[s^{-r_i}]$ with all factors proper.)

80 Fact. Let $D \in \mathbb{R}[s]^{\nu\times\nu}$ be nonsingular and r.c.r. with row powers $r_i \geq 0$, $i \in \underline{\nu}$, and column powers $k_j \geq 0$, $j \in \underline{\nu}$.

81 Let $N_\ell \in \mathbb{R}[s]^{\nu\times n_i}$ s.t. $\partial_{ri}[N_\ell] \leq r_i$ for all $i \in \underline{\nu}$.

82 Let $N_r \in \mathbb{R}[s]^{n_o\times\nu}$ s.t. $\partial_{cj}[N_r] \leq k_j$ for all $j \in \underline{\nu}$.

U.t.c.

(a) $N_rD^{-1} \in \mathbb{R}_p(s)^{n_o\times\nu}$;

(b) $D^{-1}N_\ell \in \mathbb{R}_p(s)^{\nu\times n_i}$;

(c) $N_r D^{-1} N_\ell \in \mathbb{R}_p(s)^{n_o\times n_i}$.

<u>Proof</u>. (a): Observe that

$$N_r(s)\, D(s)^{-1}$$
$$= (N_r(s)\,\mathrm{diag}[s^{-k_j}])\left(\mathrm{diag}[s^{-r_i}]\, D(s)\,\mathrm{diag}[s^{-k_j}]\right)^{-1}(\mathrm{diag}[s^{-r_i}]).$$

Now as $s \to \infty$, all terms on the RHS tend to finite constant matrices because of (82), the fact that D is r.c.r. and (61) and because $r_i \geq 0$ for all $i \in \underline{\nu}$. As a consequence the RHS tends to a finite constant matrix as $s \to \infty$ and so does the LHS. Hence $N_r\, D^{-1} \in \mathbb{R}_p(s)^{n_o\times\nu}$.

(b): Use a similar reasoning as for (a).

(c): Observe that

$$N_r(s)\, D(s)^{-1}\, N_\ell(s)$$
$$= (N_r(s)\,\mathrm{diag}[s^{-k_j}])\left(\mathrm{diag}[s^{-r_i}]\, D(s)\,\mathrm{diag}[s^{-k_j}]\right)^{-1}(\mathrm{diag}[s^{-r_i}]\, N_\ell(s)),$$

where all terms on the RHS tend to finite constant matrices as $s \to \infty$. ∎

We are now able to state a practical corollary to Theorem 52 for obtaining a well-formed PMD.

83 <u>Corollary</u> [Test for a well-formed PMD]. Consider a PMD $\mathcal{D} := [D, N_\ell, N_r, K]$ described by (3.2.1.1) and (3.2.1.2).

U.t.c. if

(a) D is r.c.r. with row powers $r_i \geq 0$, $i \in \underline{\nu}$ and column powers $k_j \geq 0$, $j \in \underline{\nu}$;

(b) $\partial_{ri}[N_\ell] \leq r_i$ for all $i \in \underline{\nu}$,

$\partial_{cj}[N_r] \leq k_j$ for all $j \in \underline{\nu}$;

(c) $K \in \mathbb{R}^{n_o\times n_i}$;

then the PMD $\mathcal{D}$ is well formed. ∎

84 <u>Comments</u>. (a) Notice that once D is r.c.r., the conditions of Corollary 83 can be checked by inspection of the <u>entries</u> of the system-matrix (51). (b) The conditions of Corollary 83 are <u>almost</u> necessary for well-formedness: see

Comments 3.4.45 and Corollary 3.4.46 below.

85 Proof of Corollary 83. We check that condition (55) is satisfied; viz., D^{-1}, N_rD^{-1}, $D^{-1}N_\ell$, and $\hat{H}$ have to be proper. Now condition (a) and Fact 78 imply that D^{-1} is proper. Moreover conditions (a) and (b) and Fact 70 imply that matrices N_rD^{-1}, $D^{-1}N_\ell$, and $N_rD^{-1}N_\ell$ are proper. Now since by condition (c) K is proper, it follows that $\hat{H} = N_rD^{-1}N_\ell + K$ is also proper. ⌘

86 Exercise. Show that the PMD $\mathcal{D} = [pI - A, B, C, D]$ satisfies the conditions of Corollary 83. ⌘

3.3.2. Exponentially Stable PMDs

We start by considering differential equations.

1 Definition. We say that the differential equation

2 $$D(p)\xi(t) = \theta_\nu \quad t \geq 0,$$

with $D(\cdot) \in \mathbb{R}[p]^{\nu\times\nu}$ nonsingular, is exponentially stable iff for all initial values of $\xi(\cdot)$ and its derivatives at $t = 0-$, (a) $\xi(\cdot)$ does not contain a distribution at $t = 0$ of the form $\sum_{i=0}^{\ell} \xi_i\delta^{(i)}(t)$, where $\xi_i \in \mathbb{R}^\nu$ for all $i \in \underline{\ell}$, and (b) $\xi(\cdot)$ is exponentially decreasing on $\mathbb{R}_+$, or equiv.

3 $$\exists\, \alpha > 0 \text{ s.t. } t \mapsto e^{\alpha t}\xi(t) \text{ is bounded on } \mathbb{R}_+.$$

Analysis. By Theorem 2.3.5.38 any solution of (2) which is C^∞ on $(0-, \infty)$ is given by

4 $$\xi(t) = \bar{\Psi}(t)x_\xi(0) \quad \forall t > 0-,$$

where with $n = \partial[\det D]$, $\bar{\Psi}(\cdot)$ is the $\nu \times n$ basis matrix for the solution space and $x_\xi(0) \in \mathbb{R}^n$ is the normalized state of (2) at $t = 0$, obtained by Algorithms 2.3.5.24 and 2.3.5.36, which use the upper triangular Hermite row form of D.

Now, condition (a) of Definition 1 is equivalent to the requirement that differential equation (2) be well formed. Hence for all initial values of $\xi(\cdot)$ and its derivatives at $t = 0-$, $\xi(\cdot)$ is given by (4), $\forall t \geq 0+$ and is at most discontinuous at t = 0 (no distribution at t = 0). Note especially that the state $x_\xi(0) = x_\xi(0+) \in \mathbb{R}^n$ is arbitrary. Therefore, (2) is exponentially stable

if and only if

5 (2) is well formed,

and

6 $\exists\ \alpha > 0$ s.t. $e^{\alpha\cdot}\bar{\Psi}(\cdot)$ is bounded on $\mathbb{R}_+$

It follows now by the construction of the basis matrix $\bar{\Psi}(\cdot)$ in Algorithm 2.3.5.24 that (6) holds if and only if

7 $$Z[\det D] \subset \mathring{\mathbb{C}}_-$$

(note that using the Hermite row form of D the Laplace transform of every element of $\bar{\Psi}(\cdot)$ is strictly proper with poles at the zeros of det D). We know also from Theorem 3.3.1.14 that (5) holds if and only if

8 $$D^{-1} \text{ is proper.}$$

Therefore, we have

9 <u>Theorem</u> [Exponentially stable differential equation]. The differential equation (2) is exponentially stable if and only if

(a)

8 $$D^{-1} \in \mathbb{R}_p(s)^{\nu\times\nu},$$

(b)

7 $$Z[\det D] \subset \mathring{\mathbb{C}}_-.$$ ∎

10 <u>Comment</u>. Note that condition (7)-(8) may be expressed as: <u>the polynomial matrix D</u> (considered as a rational matrix) <u>has no zeros in $\mathbb{C}_+$ and at ∞</u>, or equiv., <u>the rational matrix D^{-1} has no poles in $\mathbb{C}_+$ and at ∞</u>.

We consider now a PMD $\mathcal{D} := [D, N_\ell, N_r, K]$, given by (3.2.1.1)-(3.2.1.2), and recall that in (3.2.1.9) any z-i p-s trajectory $\xi(\cdot)$ of $\mathcal{D}$ is a solution of differential equation (2) which is C^∞ on (0-, ∞) and given by

12 $$\xi(t) = \Psi(t)\, P\, x_{\xi}(0) \quad \forall t > 0-.$$

Here $\Psi(\cdot)$ is any basis matrix for the C^{∞} solution space of (2), $P \in \mathbb{R}^{n\times n}$ is nonsingular, and $x_{\xi}(0) \in \mathbb{R}^{n}$ is the normalized state of $\mathcal{D}$ with $n = \partial[\det D]$.

We have then

13 Definition. The PMD $\mathcal{D} := [D, N_{\ell}, N_{r}, K]$ given by (3.2.1.1)-(3.2.1.2), is said to be exponentially stable (exp. stable) iff

(a) every z-i p-s trajectory of $\mathcal{D}$ given by (12) is exponentially decreasing on $\mathbb{R}_{+}$, or equiv.

14 $\forall x_{\xi}(0) \in \mathbb{R}^{n}$, $\exists\ \alpha > 0$ s.t.

$$e^{\alpha t}\xi(t) = e^{\alpha t}\Psi(t)P\, x_{\xi}(0) \text{ is bounded on } \mathbb{R}_{+},$$

(b)

15 the PMD $\mathcal{D}$ is well formed.

16 Comment. In Definition 13 condition (14) is the natural requirement for having exponential stability when we study state-space system descriptions, i.e., a PMD $\mathcal{D} := [pI - A, B, C, D]$. Since that PMD is always well formed, no impulsive behavior can occur at $t = 0$, under the conditions of Definitions 3.3.1.25 and 3.3.1.35. For a general PMD $\mathcal{D}$ this is not always true. Therefore, we must require (15): we want a good model.

From the analysis of differential equation (2) and Theorem 3.3.1.52 we have now

18 Theorem [Exponentially stable PMDs]. Consider the PMD $\mathcal{D} := [D, N_{\ell}, N_{r}, K]$, given by (3.2.1.1)-(3.2.1.2), with characteristic polynomial $\det D$.

U.t.c.

$\mathcal{D}$ is exp. stable,

if and only if

(a)

$$Z[\det D] \subset \mathring{\mathbb{C}}_{-}$$

(b)

the rational matrices D^{-1}, $N_{r}D^{-1}$, $D^{-1}N_{\ell}$, $\hat{H} = N_{r}D^{-1}N_{\ell} + K$ are proper (or equiv. $\mathcal{D}$ is well formed). ■

An exp. stable PMD has many desirable properties.

19 Theorem. Consider a PMD $\mathcal{D} := [D, N_\ell, N_r, K]$, given by (3.2.1.1)-(3.2.1.2) with transfer function $\hat{H}$.
U.t.c, if $\mathcal{D}$ is exp. stable, then
(a) For every value of $\xi(\cdot)$ and its derivatives at $t = 0-$, the z-i p-s trajectory $\xi(\cdot)$ and the z-i response $y(\cdot)$ are exponentially decreasing on $\mathbb{R}_+$, by which we mean:

$$\exists\ \alpha > 0 \text{ s.t. } t \mapsto e^{\alpha t}\xi(t) \text{ and } t \mapsto e^{\alpha t}y(t) \text{ is bounded on } \mathbb{R}_+.$$

(b) For every value of $\xi(\cdot)$ and its derivatives at $t = 0-$ and for every input $u(\cdot)$ p. suff. diff. on $\mathbb{R}_+$ with $u^{(j)}(0-) = \theta_{n_i}$ for all $j = 0, 1, 2, \cdots$,

20 (i) if $u(\cdot) \in L^\infty$, (equiv. is bounded on $\mathbb{R}_+$), then $\xi(\cdot)$ and $y(\cdot) \in L^\infty$;

21 (ii) if $u(\cdot) \in L^\infty$ and $\lim_{t\to\infty} u(t) = u_\infty \in \mathbb{C}^{n_i}$, then
$\lim_{t\to\infty} \xi(t) = [D^{-1} N_\ell](0)u_\infty$ and $\lim_{t\to\infty} y(t) = \hat{H}(0)u_\infty$;

22 (iii) if $u(\cdot)$ is T-periodic on $\mathbb{R}_+$, then on $\mathbb{R}_+$ there exist T-periodic functions $\xi_p(\cdot)$ and $y_p(\cdot)$ such that, as $t \to \infty$, $\xi(t) \to \xi_p(t)$ and $y(t) \to y_p(t)$; in particular if $u(t) = ke^{j\omega t}$ with $\omega = 2\pi T^{-1}$ and $k \in \mathbb{C}^{n_i}$, then $\xi_p(t) = [D^{-1} N_\ell](j\omega)\ u(t)$ and $y_p(t) = \hat{H}(j\omega)\ u(t)$;

23 (iv) if $u(t) = ke^{zt}$, where $k \in \mathbb{C}^{n_i}$ and $z \in \mathbb{C}_+$, then
$\lim_{t\to\infty} \left\{\xi(t) - [D^{-1} N_\ell](z)\ u(t)\right\} = \theta_\nu$ and $\lim_{t\to\infty} \{y(t) - \hat{H}(z)\ u(t)\} = \theta_{n_o}$. ∎

24 Comment. Note that under the specifications of statement (b) of Theorem 19 the following holds: (a) property (20) guarantees that for a bounded input, the pseudo-state and output will remain bounded; (b) properties (21)-(23) guarantee that asymptotically as $t \to \infty$ the pseudo-state and output of an exp. stable PMD inherit the behavior of the input: convergence, periodicity, exponential behavior. Properties (21)-(23) are used in set point regulation, frequency-response measurement, and tracking.

25 Proof of Theorem 19. Property (a) follows by observing that for every value of $\xi(\cdot)$ and its derivatives at $t = 0-$, the z-i $\hat{\xi}$ and $\hat{y}$ are strictly proper rational functions with poles in $\mathring{\mathbb{C}}_-$.

For the properties under (b) one observes that because of (a), we are done if these properties hold for the z-s p-s trajectory $\xi(\cdot)$ and the z-s response $y(\cdot)$. Now the latter are both given as a convolution of $u(\cdot)$ by an impulse response whose Laplace transform is a proper transfer function with poles in $\mathring{\mathbb{C}}_-$. Hence because of similarity it is sufficient to consider the z-s response $y(\cdot)$ given by

25 $$y(t) = (H * u)(t) = \int_0^t H(\tau)u(t - \tau)\, d\tau,$$

where $H(t) = \mathcal{L}^{-1}[\hat{H}](t)$ and without loss of generality $\hat{H} \in \mathbb{R}_{p,o}(s)^{n_o \times n_i}$ with $P[\hat{H}] \subset \mathring{\mathbb{C}}_-$. Therefore,

26 $$\exists\, k > 0,\ \alpha > 0 \ \text{ s.t. } \ \|H(t)\| \le ke^{-\alpha t} \quad \forall t \in \mathbb{R}_+.$$

Properties (20)-(23) are now consequences of the following.

(20): With $\|f\|_\infty := \sup\{\|f(t)\| : t \in \mathbb{R}_+\}$, (25) and (26) imply that $\|y\|_\infty \le k\alpha^{-1}\|u\|_\infty < \infty$.

(21): Defining $u(t) := 0 \quad \forall t < 0$, we have that

$$y(t) = \int_0^\infty H(\tau)\, u(t - \tau)\, d\tau \quad \forall t \in \mathbb{R}_+,$$

where for all $t \in \mathbb{R}_+$ $\|H(\tau)\, u(t - \tau)\| \le k\|u\|_\infty e^{-\alpha\tau}$ and $\tau \mapsto k\|u\|_\infty e^{-\alpha\tau}$ is absolutely integrable. Hence by the Lebesgue dominated convergence theorem [Rud.1, Th. 1.34],

$$\lim_{t\to\infty} y(t) = \int_0^\infty H(\tau)[\lim_{t\to\infty} u(t - \tau)]\, d\tau = \hat{H}(0)\, u_\infty.$$

(22): Define, for all $t < 0$, $u(t) := u(t + nT)$, where n is a sufficiently large integer s.t. $t + nT > 0$. As a consequence $u(\cdot)$ is now T-periodic on $\mathbb{R}$ with the same upperbound $\|u\|_\infty$.

Consider now the function $y_p(\cdot)$ on $\mathbb{R}_+$ given by

27 $$y_p(t) := \int_0^\infty H(\tau)\, u(t - \tau)\, d\tau \quad \forall t \in \mathbb{R}_+.$$

Observe that this function is T-periodic and well defined (for the latter observe that by (26) $\|H(\tau)\, u(t-\tau)\| \leq k\|u\|_\infty e^{-\alpha\tau}$, whence the integral in the RHS of (27) converges absolutely $\forall t \in \mathbb{R}_+$). Now by (25)-(27) for all $t \in \mathbb{R}_+$

$$\|y(t) - y_p(t)\| = \|\int_t^\infty H(\tau)\, u(t-\tau)\, d\tau\| \leq k\alpha^{-1}\|u\|_\infty e^{-\alpha t}.$$

Hence

$$\lim_{t\to\infty} y(t) = y_p(t).$$

Finally, by the substitution of $u(t) = ke^{j\omega t}$ in (27), we see that

$$y_p(t) = [\int_0^\infty H(\tau)e^{-j\omega\tau}d\tau]\, ke^{j\omega t} = \hat{H}(j\omega)\, u(t) \quad \forall t \in \mathbb{R}_+.$$

(23): Note that with $u(t) = ke^{zt}$ for all $t \in \mathbb{R}_+$ and (25) $\hat{y}(s) - \hat{H}(z)\,\hat{u}(s) = (s-z)^{-1}[\hat{H}(s) - \hat{H}(z)]\,k$, a strictly proper rational function with poles in $\mathring{\mathbb{C}}_-$. ∎

28 Exercise [Tracking property]. Consider the PMD $\mathcal{D} := [D, N_\ell, N_r, K]$ given by (3.2.1.1) and (3.2.1.2). Let $\mathcal{D}$ be exp. stable.

(a) Show that, for any $\lambda \in \mathbb{C}_+$, $\forall k \in \mathbb{N}$, $\forall u_\alpha \in \mathbb{C}^{n_i}$ with $\alpha = 0, 1, \cdots, k$, for all initial values of $\xi(\cdot)$ and its derivatives at $t = 0-$, the input

$$u(t) := \sum_{\alpha=0}^{k} u_\alpha t^\alpha e^{\lambda t} \qquad t \in \mathbb{R}_+,$$

with $u^{(j)}(0-) = \theta\ \forall j = 0, 1, 2, \cdots$, produces an output $y(\cdot)$ which satisfies as $t \to \infty$

$$\{y(t) - \sum_{\alpha=0}^{k} \hat{H}^{(\alpha)}(\lambda)u_\alpha t^\alpha e^{\lambda t}\} \to \theta$$

exponentially. Note that $\hat{H}$ is the transfer function of $\mathcal{D}$.

(b) Under the conditions of (a) with $k \in \mathbb{N}$ fixed, show that $y(t) - u(t) \to \theta$ exponentially iff the Taylor expansion of $\hat{H}(s)$ about λ reads

$$\hat{H}(s) = I + \frac{(s-\lambda)^{k+1}}{(k+1)!}\hat{H}^{(k+1)}(\lambda) + \cdots;$$

i.e., the transfer function $\hat{H}(s) - I$ must have contact of order k with 0 at λ. ∎

In many applications it is convenient to consider only the transfer function $\hat{H} \in \mathbb{R}(s)^{n_o \times n_i}$ and the I/O-map, (input-output map)

30 $$y(t) = (H * u)(t),$$

where $H(t) = \mathcal{L}^{-1}[\hat{H}(\cdot)]$ is the impulse response, $u(\cdot) : \mathbb{R}_+ \to \mathbb{R}^{n_i}$ is a p. suff. diff. input and $y(\cdot)$ is the output of the system. (When $\hat{H}$ is the transfer function of a PMD, $y(\cdot)$ is the z-s response.)

31 Definition. We say that the transfer function $\hat{H} \in \mathbb{R}(s)^{n_o \times n_i}$ is exponentially stable (exp. stable) iff $\hat{H}$ is proper and $P[\hat{H}] \subset \mathring{\mathbb{C}}_-$, or equiv. $\hat{H} \in \mathcal{R}(0)^{n_o \times n_i}$.

32 Comment. Note that $\hat{H}$ is exp. stable iff $\hat{H}$ has no poles in $\mathbb{C}_+$ and at ∞; then the impulse response is exp. decreasing on $t > 0$, with at most a Dirac δ-function at $t = 0$.

The following properties are a consequence of the proof of Theorem 19.

33 Theorem. The I/O map (30) of an exp. stable transfer function $\hat{H}$ has the following properties:

34 (i) if $u(\cdot) \in L^\infty$, then $y(\cdot) \in L^\infty$;

35 (ii) if $u(\cdot) \in L^\infty$ and $\lim_{t\to\infty} u(t) = u_\infty \in \mathbb{R}^{n_i}$ then $\lim_{t\to\infty} y(t) = \hat{H}(0)u_\infty$;

36 (iii) if $u(\cdot)$ is T-periodic on $\mathbb{R}_+$, then there exists a T-periodic function $y_p(\cdot)$ on $\mathbb{R}_+$ s.t. $\lim_{t\to\infty} y(t) = y_p(t)$; in particular for $u(t) = ke^{j\omega t}$ with $k \in \mathbb{C}^{n_i}$ and $\omega := 2\pi T^{-1}$, $y_p(t) = \hat{H}(j\omega)u(t)$;

37 (iv) if $u(t) = ke^{zt}$ for all $t \in \mathbb{R}_+$ with $k \in \mathbb{C}^{n_i}$ and $z \in \mathbb{C}_+$, then
$$\lim_{t\to\infty} \{y(t) - \hat{H}(z)\, u(t)\} = \theta.$$ ∎

Theorem 33 shows why exp. stable transfer functions are important. It is, however, most desirable that the properties of $y(\cdot)$ in Theorem 33 also hold for $\xi(\cdot)$ and $y(\cdot)$ under the specifications of Theorem 19; indeed, in many problems $\xi(\cdot)$ has components with physical meaning. Hence the question: When is it that both the transfer function $\hat{H}$ and the underlying PMD $\mathcal{D}$ are exp. stable?

40 Theorem [Exp. stability of transfer function and PMD]. Consider a PMD $\mathcal{D} := [D, N_\ell, N_r, K]$, given by (3.2.1.1)-(3.2.1.2) with transfer function $\hat{H} = N_r D^{-1} N_\ell + K$. Assume that $\hat{H}$ is exp. stable.

U.t.c.

41 $\hat{H}$ and $\mathcal{D}$ are both exp. stable

if and only if

(a)

42 $\mathcal{D}$ is well formed,

or equiv.

43 the rational matrices D^{-1}, $N_r D^{-1}$, $D^{-1} N_\ell$, and $\hat{H}$ are proper;

(b)

44 $\mathcal{D}$ has no unstable hidden modes, by which we mean that every z-i p-s trajectory $\xi(\cdot)$ of $\mathcal{D}$ associated with $\mathbb{C}_+$ eigenvalues of $\mathcal{D}$ must be reachable and observable,

or equiv.

45 $\mathcal{D}$ has no decoupling zeros in $\mathbb{C}_+$.

46 Comment. Note that the exp. stability of both $\hat{H}$ and the PMD is not automatically guaranteed! Note that z-i p-s trajectories associated with $\mathbb{C}_+$-eigenvalues λ of $\mathcal{D}$ are sums of exponential polynomials in t of the form (3.2.1.17), where $\lambda \in \mathbb{C}_+$.

47 Exercise. Consider a PMD $\mathcal{D} := [D, N_\ell, N_r, K]$. Show that $\mathcal{D}$ has no $\mathbb{C}_+$-decoupling zeros iff

48 $$\operatorname{rk}[D(s) \mid N_\ell(s)] = \nu \text{ and } \operatorname{rk}\begin{bmatrix} D(s) \\ \hline N_r(s) \end{bmatrix} = \nu \ \forall s \in \mathbb{C}_+.$$

49 Proof of Theorem 40. From Theorem 18 and Definition 31 we note that (41) holds if and only if

50 $\hat{H}$ is exp. stable $\Rightarrow$ $\mathcal{D}$ is exp. stable.

Claim 1: We have (42) and (44) $\Leftrightarrow$ (50).

$\Rightarrow$: By assumption $\hat{H}$ is exp. stable, $\mathcal{D}$ is well formed and has no unstable hidden modes. Assume for the purpose of contradiction that $\mathcal{D}$ is not exp. stable; then since $\mathcal{D}$ is well formed, according to Definition 13, $\mathcal{D}$ must have at least one z-i p-s trajectory $\xi(\cdot)$ which is associated with a $\mathbb{C}_+$ eigenvalue λ of $\mathcal{D}$ and by assumption it must be reachable and observable. So by Definitions 3.2.2.3 and 3.2.3.3 there exists an input $u(\cdot)$ with $\hat{u} \in \mathbb{R}[s]^{n_i}$ producing $\xi(\cdot)$ on $t > 0$, with $y(t) = N_r(p)\xi(t) \not\equiv \theta$ on $t > 0$. Note that $\hat{\xi}$ has only poles at $\lambda \in \mathbb{C}_+$ and that the strictly proper part of $\hat{y}$ is nonzero since $y(t) \not\equiv \theta$ on $t > 0$. This implies that $\hat{y}$ must have a pole at $\lambda \in \mathbb{C}_+$; otherwise, since $\hat{y}$ can have only poles at λ, $\hat{y}$ would be polynomial, and this is impossible since $\hat{y}$ has a nonzero strictly proper part. Hence we have that $\exists$ an input $\hat{u} \in \mathbb{R}[s]^{n_i}$ which produces a z-s response $\hat{y} = \hat{H}\hat{u}$ with a pole at $\lambda \in \mathbb{C}_+$. This implies that $\hat{H}$ must have a pole at $\lambda \in \mathbb{C}_+$; (otherwise, $\hat{y}$ cannot have a pole at $\lambda \in \mathbb{C}_+$). Since an exp. stable $\hat{H}$ has no $\mathbb{C}_+$ poles, we have a contradiction. Hence $\mathcal{D}$ must be exp. stable. ∎

$\Leftarrow$: By assumption (50) holds. Then (42) must hold by Definition 13. Note also that, since $\mathcal{D}$ is exp. stable, by Theorem 18, $Z[\det D] \subset \mathring{\mathbb{C}}_-$: it follows that $\mathcal{D}$ has no $\mathbb{C}_+$ eigenvalues. Therefore, (44) is trivially satisfied since there are no z-i p-s trajectories associated with $\mathbb{C}_+$ eigenvalues. ∎

Since Claim 1 holds now and (42) $\Leftrightarrow$ (43) by Theorem 3.3.1.52, we are done if we show (44) $\Leftrightarrow$ (45).

Claim 2: We have that (44) $\Leftrightarrow$ (45).

A rigorous bookkeeping of linearly independent z-i p-s trajectories $\xi(\cdot)$ of $\mathcal{D}$ associated with an eigenvalue $\lambda \in \mathbb{C}$ uses the Jordan chain relation [Goh.1], given by

51 $$\sum_{i=0}^{\ell} \frac{1}{(\ell - i)!} D^{(\ell-i)}(\lambda)\xi_i = \theta_\nu \quad \text{for } \ell = 0, 1, 2, \cdots.$$

This produces generalized eigenvectors $\xi_\ell \in \mathbb{C}^\nu$ for $\ell = 0, 1, 2, \cdots$ and generates linearly independent z-i p-s trajectories associated with $\lambda \in \mathbb{C}$ of the form

52 $$\xi_\ell(t) = \sum_{i=0}^{\ell} \xi_i \frac{t^{\ell-i}}{(\ell - i)!} e^{\lambda t} \quad \text{for } \ell = 0, 1, 2, \cdots.$$

We assume here for reasons of simplicity that the n eigenvalues λ_i of $\mathcal{D}$ are distinct. Under this assumption there exists a bijection between the n eigenvalues $\lambda_i \in \mathbb{C}$ of $\mathcal{D}$ and n eigenvectors of fixed direction $\xi_i \in \mathbb{C}^\nu$ satisfying

53 $$D(\lambda_i)\xi_i = \theta \quad \xi_i \neq \theta.$$

In this manner a basis for the space of z-i p-s trajectories $\xi(\cdot)$ is given by

54 $$\xi_i(t) = \xi_i e^{\lambda_i t} \qquad i \in \underline{n}.$$

Let now L be a g.c.ℓ.d. of (D, N_ℓ) and R a g.c.r.d. of (N_r, D), whence D can be factorized as

55 $$D = L\bar{D} \quad \text{and} \quad D = \tilde{D}R,$$

where by assumption D and all factors have distinct eigenvalues (i.e., the determinants have distinct roots). Note also by Definitions 3.2.2.33 and 3.2.3.22 that i-d zeros and o-d zeros are eigenvalues of $\mathcal{D}$ that are eigenvalues of L and R, resp. Furthermore, by Theorems 3.2.2.10 and 3.2.3.10 we have that (44) is equivalent to the fact that

56 $$\forall\lambda_i \in \mathbb{C}_+ \quad \bar{D}(p)\xi_i(t) = \theta \quad \forall t \geq 0 \text{ and } R(p)\xi_i(t) \not\equiv \theta \text{ on } t > 0,$$

where $\xi_i(\cdot)$, $\bar{D}$ and R are given by (53)-(55). Using this information and the assumption of distinct eigenvalues, it follows that (56) is equivalent to

$$\forall\lambda_i \in \mathbb{C}_+ \ \det L(\lambda_i) \neq 0 \text{ and } \det R(\lambda_i) \neq 0.$$

Hence we have shown that (44) ⇔ (45), and Claim 2 is proved.

End of Proof

57 Important Remark. In many applications we will encounter exponentially stable transfer functions. It is of crucial importance that the underlying PMD also be exp. stable. Consequently, we shall implicitly assume that the conditions of Theorem 40 are satisfied or we will check them by analysis (see, e.g., the plant assumption in (8.2.27); see also Fact 4.2.77 and Theorem 4.2.84).

3.4. Transfer Functions: Right-Left Fractions; Internally Proper Fractions

The objective of this section is to describe transfer functions which are not necessarily right or left fractions and similar to $C(sI - A)^{-1}B$ as obtained in state space models.

1 Definitions. Let $H \in \mathbb{R}(s)^{n_o \times n_i}$. We say that the triple of polynomial matrices $(N_r, D, N_\ell) \in E(\mathbb{R}[s])$ is right-left-coprime fraction (r.ℓ.c.f.) of H iff

(a) $\det D(s) \not\equiv 0$,

(b) $H = N_r D^{-1} N_\ell$,

(c) (N_r, D) is r.c. and (D, N_ℓ) is ℓ.c.

If we do not require (c), we say that (N_r, D, N_ℓ) is a right-left fraction (r.ℓ.f.) of H.

2 Comment. Note that r.ℓ.f.'s are naturally associated with a PMD $[D, N_\ell, N_r, 0]$, while a r.ℓ.c.f. is associated with such a minimal PMD. This association with PMDs allows to state the following result, whose proof is contained in that of Theorem 3.2.4.6.

3 Theorem [Poles of a r.ℓ.f.]. Let $H \in \mathbb{R}(s)^{n_o \times n_i}$ have a r.ℓ.f. (N_r, D, N_ℓ).

U.t.c.

(a)

4 $$p \in P[H] \Rightarrow \det D(p) = 0.$$

(b) If (N_r, D, N_ℓ) is a r.ℓ.c.f., then

5 $$p \in P[H] \Leftrightarrow \det D(p) = 0.$$ ∎

In view of Theorem 3 and Definition 3.3.2.31, we also have

8 Theorem [Exp. stable r.ℓ.c.f.s]. Let $H \in \mathbb{R}_p(s)^{n_o \times n_i}$ have a r.ℓ.c.f. (N_r, D, N_ℓ); then

$$H \text{ is exp. stable} \Leftrightarrow Z[\det D] \subset \mathring{\mathbb{C}}_-.$$ ∎

10 Comment. A PMD $\mathcal{D} := [D, N_\ell, N_r, 0]$ gives rise to a transfer function $H = N_r D^{-1} N_\ell$ which is a r.ℓ.f. If the PMD is minimal, then this r.ℓ.f. has to be coprime. If the PMD is well formed, then according to Theorem 3.3.1.52, it is not sufficient for H to be proper. Hence the following definition, which makes the association of a r.ℓ.f. with a well-formed PMD automatic.

11 Definition. We say that the r.ℓ.f. (N_r, D, N_ℓ) of $H \in \mathbb{R}(s)^{n_o \times n_i}$ is internally proper (int. pr.) iff D^{-1}, $N_r D^{-1}$, $D^{-1} N_\ell$, and $H = N_r D^{-1} N_\ell$ are proper rational matrices.

12 Fact. (a) A ℓ.f. (D_ℓ, N_ℓ) of $H \in \mathbb{R}(s)^{n_o \times n_i}$ is int. pr. iff D^{-1} and $H = D^{-1} N_\ell$ are proper rational matrices.
(b) A r.f. (N_r, D_r) of $H \in \mathbb{R}(s)^{n_o \times n_i}$ is int. pr. iff D_r^{-1} and $H = N_r D_r^{-1}$ are proper rational matrices.

Proof. A ℓ.f. (D_ℓ, N_ℓ) is r.ℓ.f. (I, D_ℓ, N_ℓ) and a r.f. (N_r, D_r) is a r.ℓ.f. (N_r, D_r, I).

We shall now characterize int. pr. right-left-fractions. For this we need the following definition and preliminary results.

14 Definition. Let $H \in \mathbb{R}(s)^{n_o \times n_i}$. We say that the triple of polynomial matrices $(D_\ell, N, D_r) \in E(\mathbb{R}[s])$ is a left-right-coprime fraction, (ℓ.r.c.f.), of H, iff

(a) $\det D_\ell(s) \not\equiv 0$, $\det D_r(s) \not\equiv 0$,

(b) $H = D_\ell^{-1} N D_r^{-1}$,

(c) (D_ℓ, N) is ℓ.c. and (N, D_r) is r.c.

If we do not require (c), we say that (D_ℓ, N, D_r) is a left-right fraction (ℓ.r.f.) of H.

15 Fact [zeros of ℓ.r.c.f.'s]. Let (D_ℓ, N, D_r) be a ℓ.r.c.f. of $H \in \mathbb{R}(s)^{n \times n}$; then $z \in \mathbb{C}$ is a zero of H iff $\det N(z) = 0$.

Proof. $H = D_\ell^{-1} N D_r^{-1}$ has a zero at $z \in \mathbb{C}$ iff $H^{-1} = D_r N^{-1} D_\ell$ has a pole at $z \in \mathbb{C}$. Now (D_r, N, D_ℓ) is r.ℓ.c.f. of H^{-1}. Hence the fact follows by Theorem 3. ∎

16 Theorem [D^{-1} proper]. Let $D \in \mathbb{R}[s]^{\nu\times\nu}$ be nonsingular.
U.t.c.

(a) $$D^{-1} \in \mathbb{R}_p(s)^{\nu\times\nu}$$

iff

17 $$D = D_{cr}\, D_-\, D_{rr},$$

where $D_{cr} \in \mathbb{R}[s]^{\nu\times\nu}$ is column-reduced, $D_- \in \mathbb{R}(s)^{\nu\times\nu}$ is biproper, and $D_{rr} \in \mathbb{R}[s]^{\nu\times\nu}$ is row-reduced.

(b) $$D^{-1} \in \mathbb{R}_p(s)^{\nu\times\nu}$$

iff

$\exists$ unimodular matrices L and $R \in \mathcal{E}(\mathbb{R}[s])$, obtained by e.o.'s over $\mathbb{R}[s]$, s.t.

18 (i) LDR is r.c.r. with row powers r_i, $i \in \underline{\nu}$, and column powers k_j, $j \in \underline{\nu}$.
(ii) $\partial_{ri}[L] \le r_i \quad \forall i \in \underline{\nu}$ and $\partial_{cj}[R] \le k_j \quad \forall j \in \underline{\nu}$.

19 Comment. Characterization (b) shows that a r.c.r. matrix is a standard form for a matrix $D \in \mathbb{R}[s]^{\nu\times\nu}$ s.t. D^{-1} is proper, modulo "economical" e.o.'s.

21 Proof of Theorem 16. (a) Setting $s = \lambda^{-1} \in \mathbb{C}$, we have that

22 $D(s)^{-1}$ is proper iff $D(\lambda^{-1}) \in \mathbb{R}_p(s)^{\nu\times\nu}$ has no zero at $\lambda = 0$.
Now a ℓ.r.c.f. of $D(\lambda^{-1})$ is of the form

23 $$D(\lambda^{-1}) = D_\ell(\lambda)^{-1}\check{D}(\lambda)D_r(\lambda)^{-1},$$

where all matrices on the RHS are polynomial in λ. Moreover, D_ℓ and D_r have Smith forms

24 $$D_\ell(\lambda) = L_\ell(\lambda)\ \mathrm{diag}[\lambda^{\gamma_j}]R_\ell(\lambda),$$

25 $$D_r(\lambda) = L_r(\lambda)\ \mathrm{diag}[\lambda^{\rho_i}]R_r(\lambda),$$

where L_r, L_ℓ, R_r, and R_ℓ are unimodular (note that $D(\lambda^{-1})$ has only poles at $\lambda = 0$). Hence by (22) and Fact 15,

26 $$D(s)^{-1} \text{ is proper iff } \det \check{D}(0) \neq 0,$$

or setting $s = \lambda^{-1}$

27 $D(s)^{-1}$ is proper iff $\check{D}(s^{-1})$ is biproper.

Notice that by setting $s = \lambda^{-1}$ in (24)-(25)

28 $D_\ell(s^{-1})$ and $D_r(s^{-1})$ are proper.

Claim 1. We have that

29 $$D_\ell(s^{-1}) = \tilde{L}(s)\, D_{cr}(s)^{-1} \text{ and } D_r(s^{-1}) = D_{rr}(s)^{-1}\tilde{R}(s)$$

where matrices L and R are biproper and matrices D_{cr} and D_{rr} are polynomial matrices which are column-reduced and row-reduced resp. Because of the symmetry in formulas (24)-(25) we shall only prove (29) for D_ℓ.

For this purpose, set $s = \lambda^{-1}$ in (24): we have

30 $$D_\ell(s^{-1}) = L_\ell(s^{-1})\left\{R_\ell(s^{-1})^{-1}\operatorname{diag}[s^{\gamma_i}]\right\}^{-1},$$

where matrices $L_\ell(s^{-1})$ and $R_\ell(s^{-1})^{-1}$ are biproper. It follows now that

31 $$R_\ell(s^{-1})^{-1} \operatorname{diag}[s^{\gamma_i}] = D_{csp}(s) + D_{cr}(s),$$

where $D_{csp} \in \mathbb{R}(s)^{\nu\times\nu}$ is strictly proper and $D_{cr} \in \mathbb{R}[s]^{\nu\times\nu}$ is c.r. Hence by (30) and (31)

$$D_\ell(s^{-1}) = L_\ell(s^{-1})[I + D_{cr}(s)^{-1}D_{csp}(s)]^{-1}D_{cr}(s)^{-1}$$
$$=: \tilde{L}(s)\, D_{cr}(s)^{-1},$$

where $\tilde{L}$ is biproper. Hence (29) is proved for D_ℓ and by the noted symmetry in (24)-(25) Claim 1 follows. ∎

Now using (27), (23), and (29) with $s = \lambda^{-1}$, it follows by

$$D_{cr}(s)^{-1}\, D(s)\, D_{rr}(s)^{-1} = \tilde{L}(s)^{-1}\, \check{D}(s^{-1})\, \tilde{R}(s)^{-1},$$

where $\tilde{L}$ and $\tilde{R}$ are biproper, that

$$D(s)^{-1} \text{ is proper iff } D_- := D_{cr}^{-1}\, D\, D_{rr}^{-1} \text{ is biproper.}$$

This establishes criterion (17).
(b): Notice in criterion (17) that D_{cr}^{-1} and D_{rr}^{-1} are proper.
Hence $\exists$ unimodular matrices L and R s.t.

32 $\quad LD_{cr} = \bar{D}_{rr} \quad$ and $\quad D_{rr}R = \bar{D}_{cr}$

with

$\bar{D}_{rr}$ r.r. with row degrees r_i,

$\bar{D}_{cr}$ c.r. with column degrees k_j,

33 $\quad \partial_{ri}[L] \le r_i \quad \forall i \in \underline{\nu}$,

34 $\quad \partial_{cj}[R] \le k_j \quad \forall j \in \underline{\nu}$.

As a consequence by (17) and (32), $LDR = \bar{D}_{rr}\, D_-\, \bar{D}_{cr}$; s.t. by criterion (3.3.1.64)

35 LDR is r.c.r. with row powers r_i and column powers k_j.

Hence by (33)-(35) the necessity of criterion (18) follows. For sufficiency note that by criterion (3.3.1.63)

$$D^{-1} = \{R(\mathrm{diag}[s^{k_j}])^{-1}\}\, D_-^{-1}\{(\mathrm{diag}[s^{r_i}])^{-1}L\},$$

where all factors on the RHS are proper.

End of Proof

We are now ready to characterize int. pr. r.ℓ.f.'s. In all results below we stress the relation with well-formed PMDs (see Definition 11).

40 <u>Theorem</u> [Int. proper r.ℓ.f.'s]. Let $H \in \mathbb{R}(s)^{n_o \times n_i}$ have a r.ℓ.f. (N_r, D, N_ℓ), where $D \in \mathbb{R}[s]^{\nu\times\nu}$.

U.t.c.

41 $\quad$ the r.ℓ.f. (N_r, D, N_ℓ) of H is <u>internally proper</u> or equiv.

42 $\quad$ the PMD $\mathcal{D} = [D, N_\ell, N_r, 0]$ is <u>well formed</u>,

(a) if and only if

43 $$D = D_{cr} D_- D_{rr},$$

where $D_- \in \mathbb{R}(s)^{\nu\times\nu}$ is biproper and D_{cr} and D_{rr} are polynomial matrices, with D_{cr} column-reduced and D_{rr} row-reduced, s.t.

$$N_r D_{rr}^{-1} \text{ and } D_{cr}^{-1} N_\ell \text{ are proper rational matrices;}$$

(b) or equivalently,

if and only if

$\exists$ unimodular matrices L and $R \in E(\mathbb{R}[s])$ s.t.

44 (i) LDR is r.c.r. with row powers r_i, $i \in \underline{\nu}$, and column powers k_j, $j \in \underline{\nu}$,

(ii) $\partial_{ri}[L] \le r_i$ $\forall i \in \underline{\nu}$ and $\partial_{cj}[R] \le k_j$ $\forall j \in \underline{\nu}$,

(iii) $\partial_{ri}[LN_\ell] \le r_i$ $\forall i \in \underline{\nu}$ and $\partial_{cj}[N_r R] \le k_j$ $\forall j \in \underline{\nu}$.

45 <u>Comments</u>. (a) Notice that criterion (17) resp. (18) is contained in criterion (43) resp. (44): this is caused by the fact that by Definition 11 D^{-1} has to be proper: the additional conditions are present to ensure that $N_r D^{-1}$, $D^{-1} N_\ell$ and $\hat{H} = N_r D^{-1} N_\ell$ are proper.

(b) Criterion (44) shows that the conditions of Corollary 3.3.1.83 for guaranteeing that a PMD $\mathcal{D} = [D, N_\ell, N_r, K]$ with K constant were almost necessary: what is needed are appropriate e.o.'s on N_r, D, N_ℓ. Indeed, we have

46 <u>Corollary</u>. [Well-formed PMD's with K constant]. Consider the PMD $\mathcal{D} = [D, N_\ell, N_r, K]$, given by (3.2.1.1)-(3.2.1.2), where $K \in \mathbb{R}^{n_o \times n_i}$ U.t.c. $\mathcal{D}$ is well formed iff criterion (43) or criterion (44) holds. ∎

50 <u>Proof of Theorem 40</u>. According to Theorem 3.3.1.52 and Definition 11, (41) and (42) are equivalent.

(a): Criterion (43) is established as follows:

We have:

41 the r.ℓ.f. (N_r, D, N_ℓ) of H is internally proper

iff

with $s = \lambda^{-1}$, the square proper rational matrix $\tilde{D}(\lambda^{-1})$ given by

51 $$\tilde{D}(\lambda^{-1}) = \begin{bmatrix} I & -N_r(\lambda^{-1}) & 0 \\ 0 & D(\lambda^{-1}) & N_\ell(\lambda^{-1}) \\ 0 & 0 & I \end{bmatrix}$$

has no zero at $\lambda = 0$.

Now there exists a ℓ.r.c.f. of $\tilde{D}(\lambda^{-1})$ of the form

52 $$\tilde{D}(\lambda^{-1}) = \begin{bmatrix} I & 0 & 0 \\ 0 & D_\ell(\lambda) & 0 \\ 0 & 0 & I \end{bmatrix}^{-1} \begin{bmatrix} I & -\check{N}_r(\lambda) & 0 \\ 0 & \check{D}(\lambda) & \check{N}_\ell(\lambda) \\ 0 & 0 & I \end{bmatrix} \begin{bmatrix} I & 0 & 0 \\ 0 & D_r(\lambda) & 0 \\ 0 & 0 & I \end{bmatrix}^{-1},$$

where all matrices on the RHS of (52) are polynomial and $D_\ell(\lambda)$ and $D_r(\lambda)$ have Smith forms (24)-(25). By Fact 15 and setting $s = \lambda^{-1}$, we therefore have by (51)-(52) that

53 $$\begin{cases} (41) \text{ holds} \\ \text{iff} \\ \check{D}(s^{-1}) \text{ is biproper and } \check{N}_r(s^{-1}) \text{ and } \check{N}_\ell(s^{-1}) \text{ are proper.} \end{cases}$$

Using Claim 1 of the Proof of Theorem 16 and (51)-(53), it is then possible to establish criterion (43).

(b): Criterion (44) is established using criterion (43) where D_{rr}^{-1} and D_{cr}^{-1} are proper. Hence there exist unimodular matrices L and R, obtained by e.o.'s, s.t. (32) holds with all its properties as listed in the proof of Theorem 16. The remainder of the proof for establishing criterion (44) is left as an exercise. ∎

We conclude this section by two corollaries characterizing int. pr. left- or right-fractions. Their proof is left as an exercise.

56.L <u>Corollary</u> [Int. proper left fractions]. Let $H \in \mathbb{R}(s)^{n_o \times n_i}$ have a ℓ.f. (D_ℓ, N_ℓ).

U.t.c.

The ℓ.f. (D_ℓ, N_ℓ) of H is <u>int. pr.</u>,

or equiv.

the PMD $\mathcal{D} = [D_\ell, N_\ell, I, 0]$ is well formed

(a) if and only if

$$D_\ell = D_{cr} D_- ;$$

57.L where $D_- \in \mathbb{R}(s)^{\nu\times\nu}$ is biproper and $D_{cr} \in \mathbb{R}[s]^{\nu\times\nu}$ is column-reduced s.t.

$$D_{cr}^{-1} N_\ell \text{ is proper,}$$

(b) or equivalently

if and only if

$\exists$ a unimodular matrix L, obtained by e.r.o.'s, s.t.

(i) $L\, D_\ell$ is row-reduced with row degrees r_i,

58.L

(ii) $\partial_{ri}[L] \leq r_i \ \forall i \in \underline{n}_o$ and $\partial_{ri}[LN_\ell] \leq r_i \ \forall i \in \underline{n}_o$. ∎

56.R Corollary [Int. proper right fractions]. Let $H \in \mathbb{R}(s)^{n_o\times n_i}$ have a r.f. (N_r, D_r).
U.t.c.

The r.f. (N_r, D_r) of H is int. pr.,
or equiv.
the PMD $\mathcal{D} = [D_r, I, N_r, 0]$ is well formed,

(a)

if and only if

$$D_r = D_-\, D_{rr},$$

57.R where $D_- \in \mathbb{R}(s)^{\nu\times\nu}$ is biproper and $D_{rr} \in \mathbb{R}[s]^{\nu\times\nu}$ is row-reduced s.t.

$$N_r D_{rr}^{-1} \text{ is proper}$$

(b) or equivalently

if and only if

$\exists$ a unimodular matrix R, obtained by e.c.o.'s, s.t.

(i) $D_r R$ is column-reduced with column degrees k_j,

58.R

(ii) $\partial_{cj}[R] \leq k_j \ \forall j \in \underline{n}_i$ and $\partial_{cj}[N_r R] \leq k_j \ \forall j \in \underline{n}_i$. ∎

59 Exercise. Prove Corollaries 56.L and 56.R. (Hint: For Corollary 56.L observe that H has an int. pr. ℓ.f. (D_ℓ, N_ℓ) iff $\begin{bmatrix} D_\ell(\lambda^{-1}) & N_\ell(\lambda^{-1}) \\ 0 & I \end{bmatrix}$ has no zero at $\lambda = 0$; moreover, this matrix admits a ℓ.c.f. of the form

$$\begin{bmatrix} \tilde{D}_\ell(\lambda) & 0 \\ 0 & I \end{bmatrix}^{-1} \begin{bmatrix} \check{D}_\ell(\lambda) & \check{N}_\ell(\lambda) \\ 0 & I \end{bmatrix} \cdots).$$

Chapter 4. Interconnected Systems

4.1. Introduction

This chapter contains a systematic development of the exponential stability of an interconnection of subsystems; it is then applied to feedback systems.

Section 2 describes an interconnected system Σ consisting of a number of subsystems. The assumptions on interconnection structure and well posedness lead to the description of a feedback system in terms of transfer functions. The assumption on an underlying PMD for each subsystem leads to the description of an overall well-formed PMD with given closed-loop characteristic polynomial (4.2.69). This leads to the exponential stability of Σ in terms of the exp. stability of a PMD, (Th. 4.2.73). Finally exponential stability in terms of transfer functions is discussed (Th. 4.2.84).

Section 3 treats in detail feedback system stability. Stability Theorem 4.3.6 specifies 4 closed-loop characteristic polynomials. Corollary 4.3.26 considers the case when the compensator is given by a fraction which is not necessarily coprime.

Section 4 establishes three important properties of feedback systems: (a) the determinant of the return-difference matrix is the ratio of the closed-loop over the open-loop characteristic polynomial; (b) plant-$\mathbb{C}_+$-zeros limit the amount of loop gain a feedback system can tolerate without becoming unstable; and (c) feedback cannot remove the plant-$\mathbb{C}_+$-zeros.

4.2. Exponential Stability of an Interconnection of Subsystems

Consider an interconnected system Σ of μ given subsystems described by their transfer function matrices

1 $$G_k \in \mathbb{R}_p(s)^{n_{ok} \times n_{ik}} \qquad k = 1, 2, \cdots, \mu.$$

For example, consider the interconnection shown in Fig. 1.

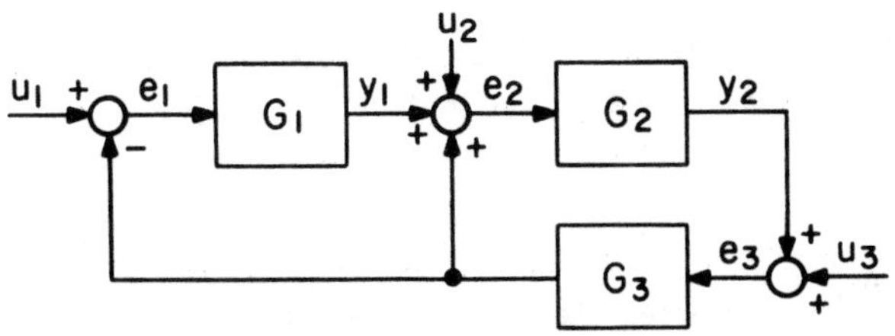

Fig. 1. An interconnected system Σ.

2 Convention. In this and the following sections we often work in the frequency domain s.t. all quantities are Laplace transforms of impulse responses and time functions on $\mathbb{R}_+$. We shall therefore omit the superscript $\hat{\ }$.

We describe now the interconnected system Σ.

3 Assumption IS. For each subsystem $G_k \in \mathbb{R}_p(s)^{n_{ok} \times n_{ik}}$, $k \in \underline{\mu}$, we have the interconnection structure shown in Fig. 2: to each subsystem G_k with input e_k and output y_k we associate a summing node with the following characteristics:
(S1) its output is the subsystem input e_k;

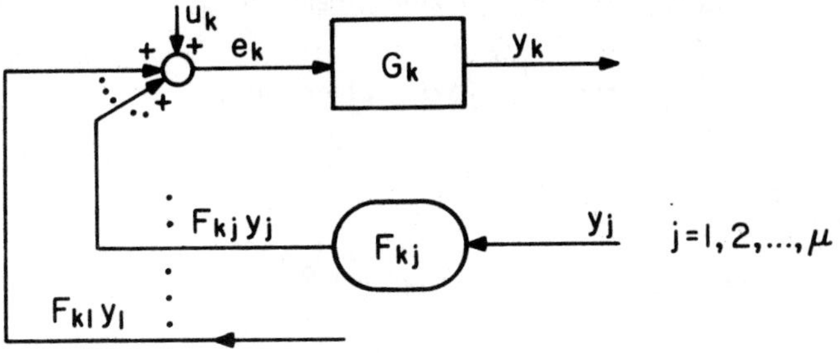

Fig. 2. Associated interconnection structure for subsystem G_k.

(S2) its inputs are:

(a) an <u>exogenous input</u> u_k (<u>always assigned</u>);

(b) other inputs which are feedbacks of the form

4 $$F_{kj}y_j \quad \text{for } j \in \underline{\mu},$$

where $F_{kj} \in \mathbb{R}^{n_{ik} \times n_{oj}}$ denotes the constant gain matrix from y_j to the kth summing node (some of these gain matrices may be zero). ∎

5 <u>Implications of Assumption IS</u>. The subsystems $G_k \in \mathbb{R}_p(s)^{n_{ok} \times n_{ik}}$, $k \in \underline{\mu}$, are interconnected according to the equations

6 $$e_k = u_k + \sum_{j=1}^{\mu} F_{kj}y_j,$$

$$\text{for } k \in \underline{\mu}.$$

7 $$y_k = G_k e_k$$

Hence by <u>aggregation</u>, i.e., by defining global quantities:

8 $$n_i := \sum_{k=1}^{\mu} n_{ik}, \quad n_o := \sum_{k=1}^{\mu} n_{ok},$$

9 $$u := [u_1^T \mid \cdots \mid u_\mu^T]^T \text{ having dimension } n_i,$$

10 $$e := [e_1^T \mid \cdots \mid e_\mu^T]^T \text{ having dimension } n_i,$$

11 $$y := [y_1^T \mid \cdots \mid y_\mu^T]^T \text{ having dimension } n_o,$$

12 $$F := [F_{kj}]_{k,j \in \underline{\mu}} \in \mathbb{R}^{n_i \times n_o},$$

13 $$G := \text{block diag}[G_k]_{k=1}^{\mu} \in \mathbb{R}_p(s)^{n_o \times n_i}.$$

Then, description (6)-(7) of interconnected system Σ is transformed into an equivalent description:

14 $$e = u + Fy,$$

15 $$y = Ge,$$

which describes the feedback system shown in Fig. 3.

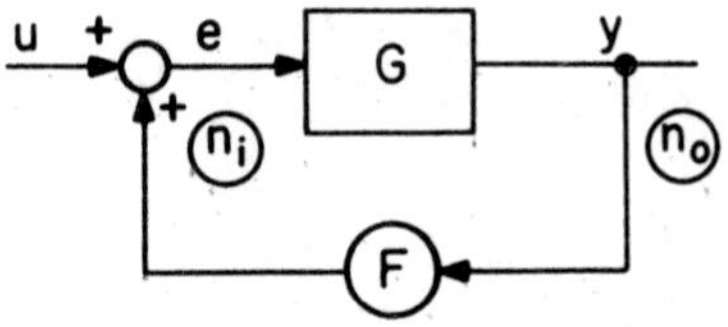

Fig. 3. Interconnected system Σ: the feedback system obtained after aggregation.

For this reason we shall call (i) u, e, y, defined by (8)-(11), the input, error, and output resp. of Σ; (ii) F, defined by (4), (8), and (12), the (feedback) gain matrix of Σ and (iii) G, defined by (1), (8), and (13), the open-loop transfer function of Σ.

For the example of Fig. 1:

$$G = \overset{\begin{matrix} n_{i1} & n_{i2} & n_{i3} \end{matrix}}{\left[\begin{array}{c|c|c} G_1 & 0 & 0 \\ \hline 0 & G_2 & 0 \\ \hline 0 & 0 & G_3 \end{array}\right]} \begin{matrix} n_{o1} \\ n_{o2}, \\ n_{o3} \end{matrix} \qquad F = \overset{\begin{matrix} n_{o1} & n_{o2} & n_{o3} \end{matrix}}{\left[\begin{array}{c|c|c} 0 & 0 & -I \\ \hline I & 0 & -I \\ \hline 0 & I & 0 \end{array}\right]} \begin{matrix} n_{i1} \\ n_{i2} \\ n_{i3} \end{matrix} .$$

■

Equations (14)-(15) and Fig. 3 show that the interconnected system Σ has the input-error and input-output transfer functions

16 $$H_{eu} : u \mapsto e, \qquad H_{yu} : u \mapsto y, \text{ resp.};$$

where

17 $$H_{eu} = (I - FG)^{-1} \in \mathbb{R}_p(s)^{n_i \times n_i}; \; H_{yu} = G(I - FG)^{-1} \in \mathbb{R}_p(s)^{n_o \times n_i}.$$

Note that

18 $$H_{eu} = I + FH_{yu}.$$

We note in Fig. 3 that all closed-loop transfer functions of Σ, i.e., from u into e, y, Fy resp. are given by H_{eu}, H_{yu}, $FH_{yu} = H_{eu} - I$.

We now introduce a well-posedness assumption.

19 Assumption WP. With F and G defined by (1), (4), (8), and (12)-(13) we assume that interconnected system Σ is s.t.

20 $$\det(I - FG)(\infty) \neq 0.$$

21 Comment. Assumption WP is a necessary and sufficient condition for all closed-loop transfer functions H_{yu}, H_{eu}, and FH_{yu} to be proper. ■

For stability purposes, in addition to Assumptions IS and WP, it is also desirable to have an assumption which permits us to represent the subsystem transfer matrices G_k as polynomial matrix fractions which are generated by subsystem PMD's $\mathcal{D}_k$ such that G_k is exp. stable if and only if $\mathcal{D}_k$ is exp. stable (see Chapter 3). We have, therefore,

25 Assumption PMD. For $k = 1, 2, \cdots, \mu$ each subsystem transfer function $G_k \in \mathbb{R}_p(s)^{n_{ok} \times n_{ik}}$ has a right-left fraction $(N_{rk}, D_k, N_{\ell k})$ generated by a PMD $\mathcal{D}_k := [D_k, N_{\ell k}, N_{rk}, 0]$ which is well formed and has no unstable hidden modes. $\forall k \in \underline{\mu}$, PMD $\mathcal{D}_k$ is described by the time-domain equations

26 $$D_k(p)\xi_k(t) = N_{\ell k}(p)e_k(t),$$

$$t \geq 0,$$

27 $$y_k(t) = N_{rk}(p)\xi_k(t)$$

where $\xi_k(\cdot)$, of dimension ν_k, is the pseudo-state of $\mathcal{D}_k$, whence in the frequency domain,

(a)

28 $$\begin{cases} (N_{rk}, D_k, N_{\ell k}) \in \mathbb{R}[s]^{n_{ok} \times \nu_k} \times \mathbb{R}[s]^{\nu_k \times \nu_k} \times \mathbb{R}[s]^{\nu_k \times n_i} \\ \text{with} \\ \quad \det D_k(s) \not\equiv 0, \end{cases}$$

(b)

29 $$G_k = N_{rk} D_k^{-1} N_{\ell k}.$$ ■

32 Fact. Assumption PMD is satisfied if and only if $\forall k = 1, 2, \cdots, \mu$, the subsystem transfer function $G_k \in \mathbb{R}(s)^{n_{ok} \times n_{ik}}$ has a r.ℓ.f. $(N_{rk}, D_k, N_{\ell k})$ (28)-(29), s.t.

33 (i) $(N_{rk}, D_k, N_{\ell k})$ is an internally proper r.ℓ.f. of G_k,

34 (ii) $\text{rk}[D_k \mid N_{\ell k}](s) = \nu_k \quad \forall s \in \mathbb{C}_+$,

35 (iii) $\text{rk}\begin{bmatrix} D_k \\ \hline N_{rk} \end{bmatrix}(s) = \nu_k \quad \forall s \in \mathbb{C}_+$. ■

Proof. According to Theorem 3.3.1.52 and Definition 3.4.11, condition (i) is necessary and sufficient for the PMD $\mathcal{D}_k$ (26)-(27) to be well formed. According to Exercise 3.3.2.47, conditions (ii) and (iii) are necessary and sufficient for the PMD $\mathcal{D}_k$ to have no $\mathbb{C}_+$-decoupling zeros. ■

36 Comment. Assumption PMD allows a left- or a right-coprime fraction for each G_k. Notice that in that case the conditions for having $\mathcal{D}_k$ well formed are simple (see Corollaries 3.4.56.L and 3.4.56.R). ■

We shall now study the properties of an interconnected system Σ under the three Assumptions (3), (19), and (25). First, let us note that by aggregation, i.e., defining

38 $$\nu = \sum_{k=1}^{\mu} \nu_k,$$

39 $$\xi := [\xi_1^T \mid \cdots \mid \xi_\mu^T]^T \text{ having dimension } \nu,$$

40 $$D := \text{block diag}[D_1, \cdots, D_\mu] \in \mathbb{R}[s]^{\nu \times \nu},$$

41 $$N_\ell := \text{block diag}[N_{\ell 1}, \cdots, N_{\ell \mu}] \in \mathbb{R}[s]^{\nu \times n_i},$$

42 $$N_r := \text{block diag}[N_{r1}, \cdots, N_{r\mu}] \in \mathbb{R}[s]^{n_o \times \nu},$$

it follows that, by (8)-(13), (26)-(29), and (38)-(42), the open-loop transfer function G of Σ, given by (13), has a r.ℓ.f. (N_r, D, N_ℓ), i.e.,

43 $$G = N_r D^{-1} N_\ell,$$

which is generated by the <u>open-loop PMD</u> $\mathcal{D}_{ye} := [D, N_\ell, N_r, 0]$ <u>of Σ</u> described in the time domain by

44 $$D(p)\xi(t) = N_\ell(p)e(t)$$

$$t \geq 0.$$

45 $$y(t) = N_r(p)\xi(t)$$

Moreover, we have

46 <u>Fact</u>. Under Assumptions (3), (19), and (25) the open-loop PMD $\mathcal{D}_{ye}$ of Σ given by (44)-(45) is well formed and has no unstable hidden modes.

<u>Proof</u>. From (38)-(43) and (13) and Fact 32 it follows that D^{-1}, $N_r D^{-1}$, $D^{-1}N_\ell$, and $G = N_r D^{-1} N_\ell$ are proper rational matrices. Hence, by Theorem 3.3.1.52, $\mathcal{D} := [D, N_\ell, N_r, 0]$ is well formed. Moreover, by (38)-(43) and Fact 32, $\text{rk}[D \mid N_\ell](s) = \nu$, $\forall s \in \mathbb{C}_+$ and $\text{rk}\begin{bmatrix} D \\ N_r \end{bmatrix}(s) = \nu$, $\forall s \in \mathbb{C}_+$. Hence, by Exercise 3.3.2.47, $\mathcal{D} := [D, N_\ell, N_r, 0]$ has no $\mathbb{C}_+$-decoupling zeros. ∎

From equations (14)-(15) and (43)-(45) we obtain a <u>closed-loop input-output PMD</u> $\mathcal{D}_{yu} := [D_g, N_\ell, N_r, 0]$ of Σ described in the time domain by

50 $$D_g(p)\xi(t) = N_\ell(p)u(t)$$

$$t \geq 0,$$

51 $$y(t) = N_r(p)\xi(t)$$

where in the frequency domain

52 $$D_g := D - N_\ell F N_r \in \mathbb{R}[s]^{\nu \times \nu}.$$

Moreover, we also obtain a closed-loop input-error PMD $\mathcal{D}_{eu} := [D_g, N_\ell, FN_r, I]$ of Σ described in the time domain by

53 $$D_g(p)\xi(t) = N_\ell(p)u(t)$$

$$t \geq 0,$$

54 $$e(t) = FN_r(p)\xi(t) + u(t)$$

where D_g is given by (52).

The transfer functions of the PMDs $\mathcal{D}_{yu}$ and $\mathcal{D}_{eu}$ are the closed-loop transfer functions H_{yu} and H_{eu} of Σ given by (16)-(18); indeed, in the frequency domain we have

55 $$H_{yu} = N_r D_g^{-1} N_\ell \in \mathbb{R}(s)^{n_o \times n_i}$$

and

56 $$H_{eu} = FN_r D_g^{-1} N_\ell + I \in \mathbb{R}(s)^{n_i \times n_i}.$$

We now have

57 Fact. Under Assumptions (3), (19), and (25) the closed-loop PMDs $\mathcal{D}_{yu}$ and $\mathcal{D}_{eu}$ are well formed.

Proof. (a) We show that $D_g \in \mathbb{R}[s]^{\nu\times\nu}$, given by (52), has a proper inverse. By Fact 46 we have that G has an int. pr. r.ℓ.f. (N_r, D, N_ℓ) whence by Theorem 3.4.40

58 $$D = D_{cr} D_- D_{rr},$$

where $D_- \in \mathbb{R}(s)^{\nu\times\nu}$ is biproper and D_{cr} and D_{rr} are polynomial matrices with D_{cr} column-reduced and D_{rr} row-reduced s.t.

59 $N_r D_{rr}^{-1}$ and $D_{cr}^{-1} N_\ell$ are proper rational matrices.

Hence by (52) and (58)-(59),

60 $$\begin{aligned} D_g^{-1} &= (D_{cr}D_-D_{rr} - N_\ell F N_r)^{-1} \\ &= D_{rr}^{-1}(D_- - D_{cr}^{-1}N_\ell F N_r D_{rr}^{-1})^{-1} D_{cr}^{-1}, \end{aligned}$$

where D_{rr}^{-1} and D_{cr}^{-1} are proper. Moreover,

61 $D_- - (D_{cr}^{-1} N_\ell)\, F(N_r D_{rr}^{-1})$ is biproper.

Indeed, by (58)-(59) it is already proper, and, using in addition (43) and (20),

$$\det(D_- - D_{cr}^{-1}N_\ell F N_r D_{rr}^{-1})(\infty) = \det D_-(\infty)\det(I - FG)(\infty) \neq 0.$$

Hence (61) follows. Now all RHS factors of (60) are proper whence D_g^{-1} is proper.

(b) The matrices $N_r D_g^{-1}$ and $D_g^{-1} N_\ell$ are proper: indeed, in

$$N_r D_g^{-1} = (N_r D_{rr}^{-1})(D_- - D_{cr}^{-1}N_\ell F N_r D_{rr}^{-1})^{-1} D_{cr}^{-1}$$

and

$$D_g^{-1}N_\ell = D_{rr}^{-1}(D_- - D_{cr}^{-1}N_\ell F N_r D_{rr}^{-1})^{-1}(D_{cr}^{-1}N_\ell),$$

all RHS factors are proper by (58), (59), and (61).

(c) By Assumption WP it follows that the transfer functions H_{yu} and H_{eu} are proper.

Hence in view of (a)-(c) and the fact that F is constant, it follows that the PMDs $\mathcal{D}_{yu}$ and $\mathcal{D}_{eu}$ are well formed, (Theorem 3.3.1.52). ∎

64 <u>Characteristic Polynomial and Exponential Stability of Interconnected System</u> Σ. Under Assumptions IS, WP, and PMD, by equations (14)-(15) and (43)-(45), and in view of Fig. 3 and Fact 57, the interconnected system Σ is a <u>well-formed</u> PMD $\mathcal{D}_g$ which maps the input u into (y, e, Fy) and is described in the time domain by

(a) the pseudo-state equation

50 $$D_g(p)\xi(t) = N_\ell(p)u(t) \qquad t \geq 0,$$

where D_g is given by (52);

(b) three readout equations given by

51 $$y(t) = N_r(p)\xi(t)$$

54 $$e(t) = FN_r(p)\xi(t) + u(t) \qquad t \geq 0.$$

65 $$Fy(t) = FN_r(p)\xi(t)$$

Notice that $\det D_g$ with $D_g \in \mathbb{R}[s]^{\nu\times\nu}$ is the characteristic polynomial of $\mathcal{D}_g$ according to Definition 3.2.1.15. Hence the following natural definitions.

66 Definitions. Consider the interconnected system Σ under Assumptions IS, WP, PMD, and its PMD $\mathcal{D}_g$ described by equations (50)-(52) and (54), (65).

(a)

67 We call the pseudo-state of Σ the pseudo-state $\xi(\cdot)$ of $\mathcal{D}_g$.

(b)

68 We call the characteristic polynomial (char. poly.) of Σ the characteristic polynomial of $\mathcal{D}_g$, namely

69 $$\chi(s) = \det D_g(s),$$

where $D_g \in \mathbb{R}[s]^{\nu \times \nu}$ with $D_g := D - N_\ell F N_r$.

(c)

70 We say that $\lambda \in \mathbb{C}$ is an eigenvalue of Σ iff λ is an eigenvalue of $\mathcal{D}_g$ or equiv., λ is a root of χ, (these λ's are often called "closed-loop eigenvalues" of Σ); we say that $\lambda \in \mathbb{C}$ is a decoupling zero of Σ iff λ is a decoupling zero of $\mathcal{D}_g$.

(d)

71 We say that Σ is exponentially stable (exp. stable) if and only if $\mathcal{D}_g$ is exp. stable. ∎

72 Comment. Observe that every definition is obtained by replacing $\mathcal{D}_g$ by Σ.

Since the PMD $\mathcal{D}_g$ is well formed, we obtain immediately by Theorem 3.3.2.18:

73 Theorem [Exp. stable interconnected system Σ]. Consider interconnected system Σ under Assumptions IS, WP, PMD and its characteristic polynomial $\chi(\cdot)$ given by (69).

U.t.c.

$$\Sigma \text{ is exp. stable}$$

if and only if

$$Z[\chi] \subset \mathring{\mathbb{C}}_-$$

or, equiv.

Σ has no $\mathbb{C}_+$-eigenvalues. ∎

74 Comment. Notice that, because Σ is a PMD $\mathcal{D}_g$ mapping u into (y, e, Fy) with pseudo-state ξ (see equations (50)-(52), (54) and (65)), exponential stability of Σ implies that $\xi(\cdot)$ and (y, e, Fy)$(\cdot)$ have properties as described in Theorem 3.3.2.19: in particular, exponentially decreasing z-i trajectories and asymptotical inheritance by (y, e, Fy)$(\cdot)$ of the properties of u$(\cdot)$ as $t \to \infty \cdots$.

75 Remark. From the analysis above it is clear that interconnected system Σ is a well-formed PMD $\mathcal{D}_g$, described by (50) - (52), (54), and (65) under Assumptions IS, WP, and

76 Assumption WF. For k = 1, 2, ⋯, μ each subsystem transfer function $G_k \in \mathbb{R}_p(s)^{n_{ok} \times n_{ik}}$ has a right-left fraction $(N_{rk}, D_k, N_{\ell k})$ generated by a well-formed PMD $\mathcal{D}_k = [D_k, N_{\ell k}, N_{rk}, 0]$ according to (26)-(29).

As a consequence Definitions 66 and Theorem 73 are valid under the Assumptions IS, WP, and WF.

This is important in the feedback compensator design problem (see below). ∎

We shall now relate the exp. stability of Σ to that of the input-output transfer function H_{yu} of Σ, given by (16)-(17). We first note that H_{yu} is the transfer function of the input-output PMD $\mathcal{D}_{yu} = [D_g, N_\ell, N_r, 0]$ of Σ and have:

77 Fact. Consider an interconnected system Σ under Assumptions IS, WP, and PMD. Let $\mathcal{D}_{yu} = [\,D_g, N_\ell, N_r, 0]$ be the input-output PMD of Σ given by (50)-(52).

U.t.c.

(a) $\mathcal{D}_{yu}$ is well formed,

(b) $\mathcal{D}_{yu}$ has no unstable hidden modes.

Proof. In view of Fact 57 we must only show that (b) holds. Observe now that the open-loop PMD $\mathcal{D}_{ye} = [D, N_\ell, N_r, 0]$ of Σ has no $\mathbb{C}_+$-decoupling zeros according to Fact 46, or equiv. by Exercise 3.3.2.47

78 $$\mathrm{rk}[D \mid N_\ell](s) = \nu \quad \forall s \in \mathbb{C}_+ \quad \text{and} \quad \mathrm{rk}\begin{bmatrix} D \\ \hline N_r \end{bmatrix}(s) = \nu \quad \forall s \in \mathbb{C}_+.$$

Now by (52),

79 $$[D_g \mid N_\ell] = [D \mid N_\ell]\begin{bmatrix} I & 0 \\ -FN_r & I \end{bmatrix} \quad \text{and} \quad \begin{bmatrix} D_g \\ \hline N_r \end{bmatrix} = \begin{bmatrix} I & -N_\ell F \\ 0 & I \end{bmatrix}\begin{bmatrix} D \\ \hline N_r \end{bmatrix},$$

where the transforming matrices in the RHS are unimodular polynomial matrices. Hence by (78)-(79)

80 $$\mathrm{rk}[D_g \mid N_\ell](s) = \nu \quad \forall s \in \mathbb{C}_+ \quad \text{and} \quad \mathrm{rk}\begin{bmatrix} D_g \\ \hline N_r \end{bmatrix}(s) = \nu \quad \forall s \in \mathbb{C}_+.$$

Therefore, by Exercise 3.3.2.47, $\mathcal{D}_{yu}$ has no $\mathbb{C}_+$-decoupling zeros. ∎

81 Exercise. Consider interconnected system Σ under Assumptions IS, WP, and PMD. Let $\mathcal{D}_{eu} = [D_g, N_\ell, FN_r, I]$ be the input-error PMD of Σ, given by (53)-(54), with transfer function H_{eu} (see equations (17) and (56)).

(a) $\mathcal{D}_{eu}$ is well formed.

(b) $\mathcal{D}_{eu}$ has no unstable hidden modes iff $\mathcal{D}_{eu}$ has no $\mathbb{C}_+$-output-decoupling zeros, or equiv.

82 $$\mathrm{rk}\begin{bmatrix} D \\ \hline FN_r \end{bmatrix}(s) = \nu \quad \forall s \in \mathbb{C}_+.$$

83 Comments. (a) In view of (78), criterion (82) is satisfied if the feedback gain matrix F of Σ (see (12)) is left invertible. (Roughly, the feedback must be "complete": Fy must determine y completely.) Note that (78) does not imply (82): e.g., take F = 0 and the open-loop PMD $\mathcal{D}_{ye} = [D, N_r, N_\ell, 0]$ not exp. stable.

(b) Criterion (82) is also satisfied if the open-loop PMD $\mathcal{D}_{ye}$ is exp. stable.

In view of Theorem 73, Fact 77, Exercise 81, and Definition 3.3.2.31 of exp. stable transfer functions, we now have by Theorem 3.3.2.40:

84 Theorem [Exp. stability of Σ by transfer functions]. Consider interconnected system Σ under Assumptions IS, WP, and PMD. Let H_{yu}, resp. $H_{eu} \in \mathcal{E}(\mathbb{R}_p(s))$, be the input-output and input-error transfer functions of Σ given by (16)-(18). Consider also the input-error PMD $\mathcal{D}_{eu} = [D_g, N_\ell, FN_r, I]$

of Σ given by (53)-(54).

U.t.c.

(a)

85 Σ is exp. stable

$\Leftrightarrow$

86 H_{yu} is exp. stable

or equiv.

87 $\mathcal{P}[H_{yu}] \subset \mathring{\mathbb{C}}_-$.

(b) Iff, in addition, $\mathcal{D}_{eu}$ has no $\mathbb{C}_+$-output-decoupling zeros, or equiv.

82 $$\mathrm{rk}\begin{bmatrix} D \\ \hline FN_r \end{bmatrix}(s) = \nu \quad \forall s \in \mathbb{C}_+,$$

then

85 Σ is exp. stable

$\Leftrightarrow$

88 H_{eu} is exp. stable

or equiv.

89 $\mathcal{P}[H_{eu}] \subset \mathring{\mathbb{C}}_-$. ∎

90 Comments. (a) Using the classification of eigenvalues of the PMD $\mathcal{D}_{yu} = [D_g, N_\ell, N_r, 0]$ of Σ, especially equation (3.2.4.18), one has that, under the conditions of Theorem 84, the characteristic polynomial of Σ (see (69)) reads

91 $$\chi(s) = \det D_g(s) = \delta(s)\pi(s),$$

where $\delta(\cdot)$ and $\pi(\cdot)$ are polynomials such that

(i) $\pi(p) = 0$ iff $p \in \mathcal{P}[H_{yu}]$,
(ii) $\delta(z) = 0$ iff $z \in \mathbb{C}$ is a decoupling zero of $\mathcal{D}_{yu}$,
(iii) $Z[\delta] \subset \mathring{\mathbb{C}}_-$.

As a consequence $Z[\chi] \subset \mathring{\mathbb{C}}_-$ iff $Z[\pi] \subset \mathring{\mathbb{C}}_-$, which explains (85)⇔(86)⇔(87).

(b) The equivalence (85)⇔(88)⇔(89) is satisfied if the gain matrix F has a left inverse in $\mathbb{R}^{n_o \times n_i}$ ("complete feedback") or if the open-loop PMD $\mathcal{D}_{ye}$ is exp. stable.

92 Corollary. Consider an interconnected system Σ under Assumptions IS and WP, with Assumption PMD strengthened to

93 Assumption PMD': The interconnected system Σ is such that $\forall k = 1, 2, \cdots, \mu$ each subsystem transfer function $G_k \in \mathbb{R}_p(s)^{n_{ok} \times n_{ik}}$ has a right-left-coprime fraction $(N_{rk}, D_k, N_{\ell k})$ generated by a well-formed and minimal PMD $\mathcal{D}_k := [D_k, N_{\ell k}, N_{rk}, 0]$ described by equations (26)-(29).

U.t.c.

(a) The open-loop PMD $\mathcal{D}_{ye} := [D, N_\ell, N_r, 0]$ and the input-output PMD $\mathcal{D}_{yu} = [D_g, N_\ell, N_r, 0]$ of Σ, given by (43)-(45) and (50)-(52), are well formed and minimal, or equiv. (N_r, D, N_ℓ) and (N_r, D_g, N_ℓ) are int. pr. r.ℓ.c.f.'s of G and H_{yu} given by (43) resp. (17).

(b) Equation (91) reads

94 $\chi(\cdot) \sim \pi(\cdot)$

or equiv. the characteristic polynomial $\chi(\cdot)$ of Σ and $\pi(\cdot)$ are equal modulo a nonzero constant s.t.

95 $p \in P[H_{yu}] \Leftrightarrow \chi(p) = 0,$

i.e., p is a pole of H_{yu} iff p is an eigenvalue of Σ. ∎

96 Exercise. Prove Corollary 92.

4.3. Feedback System Exponential Stability

Consider the feedback system Σ of Fig. 1.

1 Assumptions. In feedback system Σ of Fig. 1, P is the plant and C is the compensator s.t.

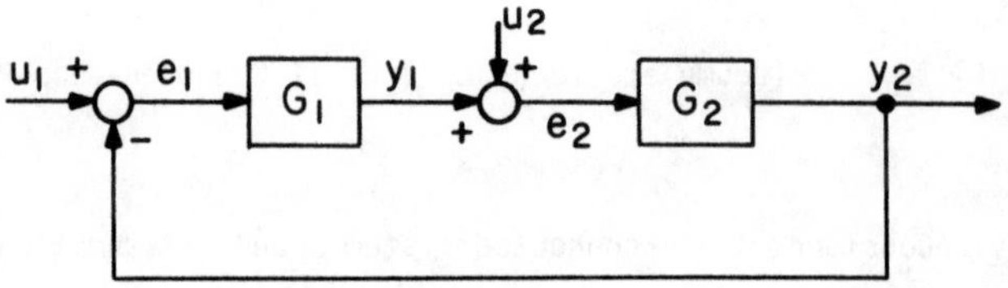

Fig. 1. The feedback system Σ under consideration.

2 $P \in \mathbb{R}_p(s)^{n_o \times n_i}$ has an int. pr. ℓ.c.f. $(D_{p\ell}, N_{p\ell})$ and an int. pr. r.c.f. (N_{pr}, D_{pr});

3 $C \in \mathbb{R}_p(s)^{n_i \times n_o}$ has an int. pr. ℓ.c.f. $(D_{c\ell}, N_{c\ell})$ and an int. pr. r.c.f. (N_{cr}, D_{cr});

4 $\det(I_{n_o} + PC)(\infty) = \det(I_{n_i} + CP)(\infty) \neq 0$.

5 Comments. (a) Assumption (4) is satisfied if P is strictly proper.

(b) In most applications Assumptions (2)-(3) are satisfied with the left denominators row-reduced and the right denominators column-reduced.

6 Theorem [Feedback system stability]. Consider feedback system Σ given by Fig. 1 under the Assumptions 1.

U.t.c.

(a) Σ is an interconnected system satisfying Assumptions IS, WP, and PMD of Sec. 4.2.

(b) The input-output transfer function H_{yu} of Σ reads

$$H_{yu} = \begin{bmatrix} H_{y_1u_1} & H_{y_1u_2} \\ H_{y_2u_1} & H_{y_2u_2} \end{bmatrix}$$

7
$$= \begin{bmatrix} C(I + PC)^{-1} & -CP(I + CP)^{-1} \\ PC(I + PC)^{-1} & P(I + CP)^{-1} \end{bmatrix}.$$

(c) The <u>characteristic polynomial</u> of Σ, defined in (4.2.69), has four <u>equivalent expressions</u> (equal modulo a nonzero constant):

8 $$\chi(s) \sim \det [D_{c\ell} D_{pr} + N_{c\ell} N_{pr}](s),$$

9 $$\sim \det [D_{p\ell} D_{cr} + N_{p\ell} N_{cr}](s),$$

10 $$\sim \det \left[\begin{array}{c|c} D_{cr} & N_{pr} \\ \hline -N_{cr} & D_{pr} \end{array}\right](s),$$

11 $$\sim \det \left[\begin{array}{c|c} D_{c\ell} & N_{c\ell} \\ \hline -N_{p\ell} & D_{p\ell} \end{array}\right](s).$$

(d)

12 $$p \in P[H_{yu}] \Leftrightarrow \chi(p) = 0.$$

(e)

13 $$\Sigma \text{ is exp. stable} \Leftrightarrow Z[\chi] \subset \mathring{\mathbb{C}}_-.$$

(f) If $H_{eu} : u = (u_1, u_2) \mapsto e = (e_1, e_2)$ is the input-error transfer function of Σ, then

14 $$p \in P[H_{yu}] \Leftrightarrow p \in P[H_{eu}].$$ ∎

15 <u>Comment</u>. The four expressions of the char. poly. of Σ reflect the four possible fractional representations of P and C in (2)-(3).

16 <u>Proof of Theorem 6</u>.

(a) Identify $G_1 := C$ and $G_2 := P$ as subsystem transfer matrices, with $n_{o1} = n_i$, $n_{i1} = n_o$, $n_{o2} = n_o$, and $n_{i2} = n_i$. From Fig. 1 it follows that Σ satisfies Assumption IS with F and G in (4.2.12)-(4.2.13) reading

17 $$F = \left[\begin{array}{c|c} 0 & -I \\ \hline I & 0 \end{array}\right], \quad G = \left[\begin{array}{c|c} C & 0 \\ \hline 0 & P \end{array}\right].$$

Assumption WP is satisfied because of (4); indeed, by (17),

$$\det[I - FG](\infty) = \det\left[\begin{array}{c|c} I & P \\ \hline -C & I \end{array}\right](\infty) = \det(I + PC)(\infty) = \det(I + CP)(\infty) \neq 0.$$

Finally, Assumption PMD is satisfied by assumptions (2)-(3). Hence Σ is an interconnected system satisfying Assumptions IS, WP, and PMD.

(b) Exercise: follows by straightforward computation.

(c) We shall compute (8) in detail. Choose for $G_1 := C$ an int. pr. ℓ.c.f. $(D_{c\ell}, N_{c\ell})$ and for $G_2 := P$ an int. pr. r.c.f. (N_{pr}, D_{pr}). Then, in (4.2.40)-(4.2.42), we have

18 $\quad D = \text{block diag}[D_{c\ell}, D_{pr}],$

19 $\quad N_\ell = \text{block diag}[N_{c\ell}, I],$

20 $\quad N_r = \text{block diag}[I, N_{pr}].$

As a consequence, (4.2.52) reads

21 $$D_g = D - N_\ell F N_r = \left[\begin{array}{c|c} D_{c\ell} & N_{c\ell}N_{pr} \\ \hline -I & D_{pr} \end{array}\right]$$

Hence in (4.2.69) the characteristic polynomial of Σ reads

$$\chi(s) = \det D_g(s) = \det\left[\begin{array}{c|c} D_{c\ell} & D_{c\ell}D_{pr} + N_{c\ell}N_{pr} \\ \hline -I & 0 \end{array}\right]$$

$$\sim \det[D_{c\ell}\, D_{pr} + N_{c\ell}\, N_{pr}](s).$$

Hence (8) holds.

Expressions (9)-(11) are computed similarly by choosing for $G_1 := C$, $G_2 := P$ resp. (a) an int. pr. r.c.f. and an int. pr. ℓ.c.f., (b) an int. pr. r.c.f. and an int. pr. r.c.f., (c) an int. pr. ℓ.c.f. and an int. pr. ℓ.c.f.. We notice now that, in the computations above, we worked under assumptions (2)-(4), whence Σ is an interconnected system satisfying Assumptions IS, WP, and PMD'(see Corollary 4.2.92). Hence

any expression on the RHS of (8)-(11) is a denominator
22 determinant of r.ℓ.c.f. (N_r, D_g, N_ℓ) of the input-output transfer function H_{yu} of Σ.

Expressions (8)-(11) are therefore equivalent for the reasons sketched below:

1) Any r.ℓ.c.f. (N_r, D_g, N_ℓ) of H_{yu} can be converted into a r.c.f. $(N_r\bar{N}_r, \bar{D}_r)$ of H_{yu}, where D_g and $\bar{D}_r$ have the same nonunity invariant polynomials. (Hint: Use generalized Bezout identities for the coprime fractions $N_r D_g^{-1}$ and $D_g^{-1} N_\ell$ and repeat the arguments in the proof of Theorem 2.4.4.12.)

2) According to Theorem 2.4.2.41, all denominators of any coprime fraction of H_{yu} have the same nonunity invariant polynomials; in particular their determinants are equivalent.

(d) This is a consequence of (22) and Theorem 3.4.3.

(e) This is a consequence of (c) and Theorem 4.2.73.

(f) Notice that the feedback matrix $F \in \mathbb{R}^{(n_o+n_i)\times(n_o+n_i)}$, given by (17), is invertible. Therefore, (14) is a consequence of (4.2.18). ∎

23 <u>Exercise</u>. Consider Σ defined in (1)-(4). Let ζ be a zero of $\chi_p = \det D_{pr}$. Let $N_{c\ell}(\zeta)$ be full rank. Show that (a) in the single-input single-output case it is always true that $\chi(\zeta) \neq 0$ (where χ is given by (8)); (b) in the multi-input multi-output case, it <u>may</u> happen that $\chi(\zeta) = 0$. (Roughly speaking, with pole-zero cancellations between P and C ruled out, (a) for the single-input single-output case, a plant pole ζ <u>can never</u> be a closed-loop eigenvalue; (b) in the multi-input multi-output case, a plant pole ζ <u>may</u> be a closed-loop eigenvalue.) ∎

In feedback compensator design we may encounter a compensator which is not a coprime fraction.

26 <u>Corollary</u> [C not a coprime fraction]. Under the assumptions of Theorem 6, where Assumption (3) is replaced by

27 $C \in \mathbb{R}_p(s)^{n_i \times n_o}$ has an int. pr. ℓ.f. $(D_{c\ell}, N_{c\ell})$ or an int. pr. r.f. (N_{cr}, D_{cr}),

we have that:

(a) Feedback system Σ, shown in Fig. 1, is an interconnected system Σ satisfying assumptions IS, WP, and WF of Sec. 4.2 (see Remark 4.2.75).
(b) Σ has input-output- and input-error transfer functions H_{yu} resp. H_{eu}, given by (7) and

28 $$H_{eu} = I + FH_{yu}$$

with

17 $$F = \left[\begin{array}{c|c} 0 & -I \\ \hline I & 0 \end{array}\right].$$

(c) The four possible choices of fractions for P and C in (2) and (27) define four well-formed closed-loop PMDs $\mathcal{D}_g$ of Σ as in (4.2.64) et seq.; each choice generates a char. poly. χ for Σ, given by an expression on the RHS of (8)-(11); furthermore, Σ is exp. stable iff $Z[\chi] \subset \mathring{\mathbb{C}}_-$.
(d) If P has an int. pr. r.c.f. (N_{cr}, D_{cr}) and C has an int. pr. ℓ.f. $(D_{c\ell}, N_{c\ell})$, then the following holds: Let $L \in E(\mathbb{R}[s])$ be any g.c.ℓ.d. of $(D_{c\ell}, N_{c\ell})$, whence

29 $$D_{c\ell} = L\bar{D}_{c\ell} \quad \text{and} \quad N_{c\ell} = L\bar{N}_{c\ell}$$

for some $\bar{D}_{c\ell}, \bar{N}_{c\ell} \in E[\mathbb{R}[s]]$ s.t.

30 $$\text{rk}[\bar{D}_{c\ell} \mid \bar{N}_{c\ell}](s) = n_i \quad \forall s \in \mathbb{C}.$$

Then the char. poly. of Σ reads

31 $$\chi(s) = \det[D_{c\ell}D_{pr} + N_{c\ell}N_{pr}](s)$$
$$= \det L(s) \det[\bar{D}_{c\ell}D_{pr} + \bar{N}_{c\ell}N_{pr}](s),$$

where

32 $$p \in P[H_{yu}] \Leftrightarrow p \in P[H_{eu}] \Leftrightarrow \det[\bar{D}_{c\ell}D_{pr} + \bar{N}_{c\ell}N_{pr}](p) = 0$$

and

33 $$z \text{ is a decoupling zero of } \Sigma \Leftrightarrow \det L(z) = 0$$
$$\Leftrightarrow z \text{ is a decoupling zero of C}.$$

34 Exercise. Prove Corollary 26. (For (33) note that since F is invertible, z is a decoupling zero of Σ iff z is a decoupling zero of the open-loop PMD $[D, N_\ell, N_r, 0]$ given in (4.2.44)-(4.2.45).)

35 Remark. In the spirit of part (c) of Corollary 26, suppose that both P and C are specified by not necessarily coprime fractions. Then it is easy to check that the characteristic polynomial of Σ is still given by the corresponding expression in the RHS of (8)-(11).

4.4. Special Properties of Feedback Systems

Consider feedback system Σ, shown in Fig. 4.3.1, under the assumptions (4.3.1) with $P = N_{pr} D_{pr}^{-1} \in E(\mathbb{R}_{p,o}(s))$ and $C = D_{c\ell}^{-1} N_{c\ell}$.

1 (a) Note that in (4.3.8), the char. poly. of Σ satisfies

$$\begin{aligned}\chi_\Sigma(s) &= \det[D_{c\ell}D_{pr} + N_{c\ell}N_{pr}](s)\\ &= \det(I + PC)(s)\cdot\det D_{c\ell}(s)\cdot\det D_{pr}(s).\end{aligned}$$

where $\det D_{c\ell} =: \chi_c$ and $\det D_{pr} =: \chi_p$ are the compensator- and plant-char. poly.'s. Hence

2 $$\det(I + PC)(s) = \chi_\Sigma(s)\Big/\big(\chi_c(s)\ \chi_p(s)\big),$$

i.e., the determinant of the return difference matrix is the ratio of the closed-loop char. poly. over the open-loop char. poly. of Σ.

This relation is useful for establishing a graphical test for stability (Nyquist criterion) (see Exercise 11 below).

3 (b) Note that if $C = k\, D_{c\ell}^{-1} N_{c\ell}$, where $k \in \mathbb{R}$ is an adjustable loop gain, then the char. poly. of Σ satisfies

4 $$\begin{aligned}\chi(s, k) &= \det[D_{c\ell}D_{pr} + kN_{c\ell}N_{pr}](s)\\ &= k^{n_i} \det[\tfrac{1}{k} D_{c\ell}D_{pr} + N_{c\ell}N_{pr}](s).\end{aligned}$$

Suppose now that k is very large; then the polynomial in (4) has coefficients which are close to the coefficients of the polynomial $\det[N_{c\ell}N_{pr}]$. Therefore, by the continuous dependence of the zeros of a polynomial on its coefficients, the char. poly. $\chi(s, k)$ has, for k large, at least one zero (say z_i) close to every zero of $\det[N_{c\ell}N_{pr}]$ (say z_i'). Furthermore, for all such zeros of $\chi(s, k)$, $|z_i - z_i'| \to 0$ as $k \to \infty$. Therefore, if the plant P has $\mathring{\mathbb{C}}_+$-zeros, then as the feedback becomes tighter (i.e., $k \to \infty$), a closed-loop eigenvalue tends to each of the $\mathring{\mathbb{C}}_+$-zeros of the plant, whence for k sufficiently large Σ will be unstable.

5 <u>Conclusion</u>. A plant with $\mathring{\mathbb{C}}_+$-zeros imposes a bound on the amount of loop gain that a feedback system can tolerate.

6 <u>Exercise</u>. Show that the statement (5) <u>may</u> be false if $\mathring{\mathbb{C}}_+$ is replaced by $\mathbb{C}_+$. (Hint: Let k_1, k_2, p_1, p_2 be positive numbers; let $p(s) = k_1 s/(s + p_1)^2$, $c(s) = k_2/(s + p_2)$. With k_1 <u>finite</u> but $k_2 \to \infty$, show that the zeros of $\chi(s, k_2)$ remain in $\mathring{\mathbb{C}}_-$. Give in addition a proof based on the Nyquist criterion.)

8 (c) In applications the <u>usual I/O map</u> of Σ is given by

9
$$H_{y_2u_1} = PC(I + PC)^{-1} = P(I + CP)^{-1}C$$
$$= N_{pr}(D_{c\ell}D_{pr} + N_{c\ell}N_{pr})^{-1}N_{c\ell}.$$

Now, if Σ is exp. stable, then, by (4.3.8), $Z[\det(D_{c\ell}D_{pr} + N_{c\ell}N_{pr})] \subset \mathring{\mathbb{C}}_-$ and there cannot be any cancellation of a right nonunimodular common factor with $\mathbb{C}_+$-eigenvalues between N_{pr} and the denominator of $H_{y_2u_1}$ in (9). Hence we have

10
$$Z[P] \cap \mathbb{C}_+ \subset Z[H_{y_2u_1}],$$

i.e., <u>feedback cannot remove the $\mathbb{C}_+$-zeros of the plant</u>.

The following exercise completes the information given at the end of (a).

11 <u>Exercise</u> [Multivariable Nyquist test]. Consider the system Σ specified by the assumptions (4.3.1).
(a) Assume that neither χ_p nor χ_c have $j\omega$-axis zeros. Let n_p (n_c) denote the number of $\mathbb{C}_+$-zeros of χ_p (χ_c, resp.). Let R be real, positive, and arbitrarily large. Define the <u>oriented</u> curve D_R be the straight-line segment from (0, -jR) to (0,jR) and the right half-plane semicircle joining these

points. Note that D_R is traversed clockwise. Let $\phi : \mathbb{C}_+ \to \mathbb{C}$ be analytic on D_R; then the Nyquist diagram of ϕ is, by definition, the oriented curve $\phi(D_R)$, namely the map of D_R by ϕ. Use the principle of the argument to show that

$Z[\chi_\Sigma(s)] \subset \mathring{\mathbb{C}}_-$ $\Leftrightarrow$ the Nyquist diagram of det[I + PC](s) does not go through the origin and encircles it $n_p + n_c$ times counterclockwise.

(c) In case χ_P or χ_C have $j\omega$-axis zeros, show that for the statement above to remain true, the contour D_R must be indented on the left (see Exercise 4.3.23).

The Nyquist criterion is the basis for the following robust stability result.

12 Remark [Robust stability] [Doy.1]. Call $\Sigma(P, C)$ the system described by (4.3.1)-(4.3.3) and $P \in \mathbb{R}_{p,o}(s)^{n_o \times n_i}$. Model plant perturbations as follows: P is replaced by (I + M)P, resulting in a perturbed system $\Sigma((I + M)P, C)$. As in (4.3.2), (I + M)P is specified by an int. proper coprime fraction, hence $\chi_{(I+M)P}$ is well defined. The size of the perturbation M is bounded by a given continuous tolerance curve $\ell(\cdot) : \mathbb{R}_+ \to \mathbb{R}_+$, s.t. $\ell(\omega) > 0$, $\forall\omega \in \mathbb{R}_+$, and $\exists k \in \mathbb{N}$ s.t. $\ell(\cdot)\omega^k > 1$ for ω sufficiently large.

We say that $M \in \mathcal{M}$ iff $M \in \mathbb{R}(s)^{n_o \times n_o}$ is s.t.

(a) $(I + M)P \in \mathbb{R}_{p,o}(s)^{n_o \times n_i}$;

(b) $\sigma_{max}[M(j\omega)] < \ell(\omega)$ $\forall\omega \in \mathbb{R}_+$, where $\sigma_{max}[\cdots]$ denotes the maximal singular value;

(c) the number of $\mathbb{C}_+$-zeros of $\chi_{(I+M)P}$ equals the number of $\mathbb{C}_+$-zeros of χ_P, counting multiplicities. (Equivalently, P and (I + M)P have the same number of $\mathbb{C}_+$-poles, with multiple poles counted according to their Mcmillan degree [Kai.1, Sec. 6.5].)

The following can then be proven:

Let $\Sigma(P, C)$ be exp. stable

U.t.c.

$$\forall M \in \mathcal{M},\ \Sigma\big((I + M)P,\ C\big) \text{ is exp. stable}$$

$\Leftrightarrow$

$$\sigma_{max}[PC(I + PC)^{-1}(j\omega)] \le 1/\ell(\omega),\ \forall\omega \in \mathbb{R}_+.$$

13 Comment. A remarkable feature of the result of Remark 12 is that the three transfer functions P, C, and (I + M)P may be unstable. In fact, by suitable choice of M, P, and (I+M)P may have no zeros and no poles in common!

14 Exercise. Let

$$P = \begin{bmatrix} \frac{1}{s-1} & \frac{s}{s^2+s+1} \\ 0 & \frac{s-2}{(s+1)(s+2)} \end{bmatrix}$$

Find an $M \in \mathbb{R}_p(s)^{2\times 2}$ such that $\mathbb{C}_+$-zero of P and the $\mathbb{C}_+$-pole of P are moved to $2 + \Delta_2$ and to $1 + \Delta_1$ respectively. ∎

Chapter 5. Single-Input Single-Output Systems

5.1. Introduction

The purpose of this chapter is to treat, in the simple context of single-input single-output feedback systems [Chen.1], a number of design problems that will be treated in the multivariable case in Chapters 6, 7, and 8.

In Sect. 2 we study, for a given plant, the problem of designing a two-input one-output compensator. Such compensator has the advantage that it allows independent adjustment of the closed-loop dynamics and of the I/O map. The realization of the compensator is carefully described because it is of crucial importance in the determination of closed-loop characteristic polynomial (5.2.26). This characteristic polynomial is used in the definition of $\mathcal{U}$-stability where the closed-loop eigenvalues in the undesirable set $\mathcal{U}$ of the complex plane are ruled out (Theorem 5.2.37). The properness of the compensator is guaranteed by simple conditions (Theorem 5.2.47). Corollary 5.2.56 describes the class of all achievable I/O maps using such a compensator and the given plant. Finally, the concept of robust asymptotic tracking is carefully defined and Theorem 5.2.71 specifies the necessary and sufficient conditions that must be satisfied so that robust asymptotic tracking is achieved.

Section 3 on design contains two algorithms. The first algorithm obtains a compensator that achieves a prescribed I/O map for the closed-loop system; this algorithm is followed by comments which justify it and which describe ideas which will be useful in the multivariable case in Sec. 6.2. The second algorithm obtains a compensator which achieves robust asymptotic tracking.

5.2. Problem Statement and Analysis

We consider a linear time-invariant feedback system Σ_2 with single input, v_1, and single output, y_2, as shown in Fig. 1, where u_1, u_2, and d_o are disturbance scalar inputs (with d_o a possible disturbance at the output of the plant p), and y_1 is the scalar output of the compensator. The problem is the following. Given a plant transfer function $p \in \mathbb{R}_{p,o}(s)$, we wish to design a <u>proper</u> compensator with <u>two inputs</u>, namely v_1 and e_1, and <u>one output</u> y_1, s.t.

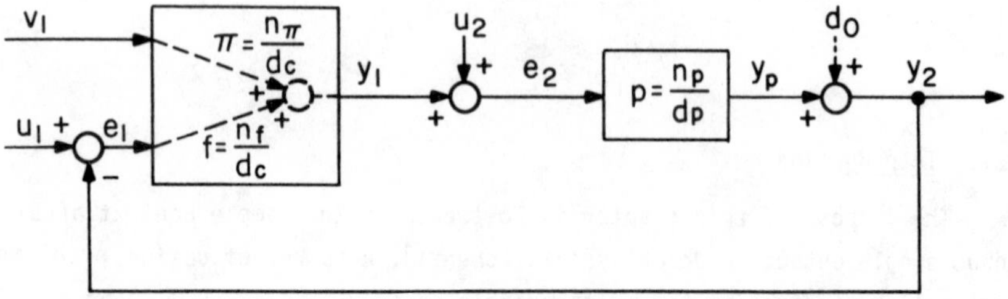

Fig. 1. The feedback system Σ_2 under consideration.

(i) system Σ_2 is stable in some sense;

(ii) the system I/O map $h_{y_2v_1}$ has, for example, a "good" step response; and

(iii) other design specifications (such as tracking, disturbance rejection, desensitization, etc.) are satisfied.

The compensator under consideration can be viewed as consisting of a precompensator with transfer function π: $v_1 \mapsto y_1$ and a feedback compensator with transfer function $f : e_1 \mapsto y_1$. In terms of realization, we let $\pi := n_\pi/d_c$ and $f := n_f/d_c$, where d_c is a <u>common denominator</u> of the rational transfer functions π and f, and n_π, $n_f \in \mathbb{R}[s]$ are the corresponding numerator, respectively. We propose to realize the two-input one-output controller using the observer canonical form [Kai.1, Fig. 2.1.9, p. 43]. More precisely, $1/d_c$ is first realized by using constant-gain feedback loops around cascaded integrators; the inputs v_1 and e_1 are then fed through appropriate constant gains to the integrator inputs to obtain n_π and n_f, respectively.

This section establishes three theorems: (a) for the <u>U-stability</u> of feedback system Σ_2, (b) for the existence of a <u>proper</u> compensator, and (c) for <u>robust asymptotic tracking</u> over a specified class of inputs, respectively.

1 <u>Assumptions</u>. For system Σ_2, shown in Fig. 1, we assume that:

(a) the plant is given by its transfer function

2 $$p = n_p/d_p \in \mathbb{R}_{p,o}(s),$$

where n_p and d_p are coprime polynomials with $\partial[d_p] \geq 1$; moreover, p is generated by the well-formed and minimal plant PMD $\mathcal{D}_p = [d_p, 1, n_p, 0]$ with plant pseudo-state $\xi_p(\cdot)$ and given by

3 $$d_p(p)\xi_p(t) = e_2(t)$$

$$t \geq 0.$$

4 $$y_2(t) = n_p(p)\xi_p(t)$$

(b) the compensator is given by the transfer function

5 $$c = [\pi \mid f] = [n_\pi \mid n_f]/d_c \in \mathbb{R}_p(s)^{1\times 2},$$

where n_π, n_f and $d_c \in \mathbb{R}[s]$; moreover, c is generated by the well-formed compensator PMD $\mathcal{D}_c = [d_c, [n_\pi \mid n_f], 1, 0]$, with compensator pseudo-state $\xi_c(\cdot)$ and given by

6 $$d_c(p)\xi_c(t) = n_\pi(p)v_1(t) + n_f(p)e_1(t)$$

$$t \geq 0.$$

7 $$y_1(t) = \xi_c(t)$$

10 <u>Analysis</u>. Under the Assumptions 1, system Σ_2, shown in Fig. 1, is described in the time domain by a PMD $\mathcal{D}_g$, with global input $u = (u_1, u_2, v_1)^T$, pseudo-state $\xi = (\xi_p, \xi_c)^T$, and output $(y^T, e^T)^T$, where $y := (y_1, y_2)^T$ and $e := (e_1, e_2)^T$, by

(a) a pseudo-state equation, viz.,

11 $$D(p)\xi(t) = N_\ell(p)u(t) \qquad t \geq 0,$$

where

12 $$D = \left[\begin{array}{c|c} d_p & -1 \\ \hline n_f n_p & d_c \end{array}\right] \in \mathbb{R}[s]^{2\times 2},$$

13 $$N_\ell = \left[\begin{array}{c|c|c} 0 & 1 & 0 \\ \hline n_f & 0 & n_\pi \end{array}\right] \in \mathbb{R}[s]^{2\times 3},$$

(b) Two readout maps, viz.,

14 $$y(t) = N_r(p)\xi(t)$$

$$t \geq 0,$$

15 $$e(t) = FN_r(p)\xi(t) + Gu(t)$$

where

16 $$e(t) = Fy(t) + Gu(t) \qquad t \geq 0,$$

with

17 $$N_r = \left[\begin{array}{c|c} 0 & 1 \\ \hline n_p & 0 \end{array}\right] \in \mathbb{R}[s]^{2\times 2},$$

18 $$F = \left[\begin{array}{c|c} 0 & -1 \\ \hline 1 & 0 \end{array}\right] \in \mathbb{R}^{2\times 2}, \quad G = \left[\begin{array}{c|c|c} 1 & 0 & 0 \\ \hline 0 & 1 & 0 \end{array}\right] \in \mathbb{R}^{2\times 3}.$$

Moreover, d_o, disturbance at the output, is dynamically accounted for by replacing y_2 and u_1 resp. by $y_2 - d_o$ and $u_1 - d_o$ in the equations (11)-(18) describing the PMD $\mathcal{D}_g$ of Σ_2.

Note also that the PMD $\mathcal{D}_g$ of Σ_2 contains the global input-output PMD

19 $$\mathcal{D}_{yu} = [D, N_\ell, N_r, 0]$$

and the global input-error PMD

20 $$\mathcal{D}_{eu} = [D, N_\ell, FN_r, G],$$

with transfer functions

21 $$H_{yu} = N_r D^{-1} N_\ell \in \mathbb{R}_p(s)^{2\times 3}$$

22 $$H_{eu} = FN_r D^{-1} N_\ell + G \in \mathbb{R}_p(s)^{2\times 3},$$

where, because $F^{-1} = -F$ (see (18)),

23 $$H_{eu} = FH_{yu} + G \text{ and } H_{yu} = -FH_{eu} + FG.$$ ■

The following fact is not unexpected in view of Fact 4.2.57.

25 <u>Fact</u>. Under the Assumptions 1, feedback system Σ_2 has a well-formed PMD $\mathcal{D}_g$ (11)-(18) and the same holds for the PMDs $\mathcal{D}_{yu}$ and $\mathcal{D}_{eu}$ described in (19)-(20).

<u>Proof</u>. In view of Theorem 3.3.1.52 and the descriptions of $\mathcal{D}_g$, $\mathcal{D}_{yu}$, and $\mathcal{D}_{eu}$ it is sufficient to show that under the Assumptions 1,

$$D^{-1},\ N_r D^{-1},\ D^{-1} N_\ell, \text{ and } H_{yu} = N_r D^{-1} N_\ell \in E(\mathbb{R}_p(s)),$$

where D, N_r, and N_ℓ are given by resp. (12), (13), and (17). Now this follows easily by inspection: for example,

$$D^{-1} = (d_c\, d_p + n_f\, n_p)^{-1} \left[\begin{array}{c|c} d_c & 1 \\ \hline -n_f n_p & d_p \end{array}\right] = (1 + fp)^{-1} \left[\begin{array}{c|c} d_p^{-1} & d_c^{-1} d_p^{-1} \\ \hline -fp & d_c^{-1} \end{array}\right],$$

where <u>all</u> elements in the RHS of the second equality are proper because of assumptions (2) and (5). ∎

Note also from the analysis done in (10) that the PMDs $\mathcal{D}_g$, $\mathcal{D}_{yu}$, and $\mathcal{D}_{eu}$ of Σ_2 have a common pseudo-state ξ and a common characteristic polynomial

26 $$\chi(s) = \det D(s) = (d_c\, d_p + n_f\, n_p)(s),$$

which in view of Fact 25 and Theorem 3.3.2.18 must have roots in $\mathring{\mathbb{C}}_-$ for $\mathcal{D}_g$, $\mathcal{D}_{yu}$, and $\mathcal{D}_{eu}$ to be exp. stable. Hence the following definitions make sense.

27 <u>Definitions</u>. Consider feedback system Σ_2, shown in Fig. 1, under the Assumptions 1. Consider also the PMD $\mathcal{D}_g$ of Σ described by (11)-(18).

(a)

28 We call <u>pseudo-state of Σ_2</u> the pseudo-state $\xi = (\xi_p, \xi_c)^T$ of $\mathcal{D}_g$.

(b)

29 We call <u>char. poly. of Σ_2</u> the char. poly. of $\mathcal{D}_g$, namely,

26 $$\chi(s) = \det D(s) = (d_c\, d_p + n_f\, n_p)(s).$$

(c)

30 We say that <u>$\lambda \in \mathbb{C}$</u> is a (closed-loop) <u>eigenvalue of</u> Σ_2 iff λ is an

eigenvalue of $\mathcal{D}_g$ or equiv. λ is a root of χ.

(d)

31 We say that Σ_2 <u>is exp. stable</u> iff $\mathcal{D}_g$ is exp. stable.

From the definitions above, Fact 25 and Theorem 3.3.2.18, we now have

32 <u>Fact</u>. Consider feedback system Σ_2, shown in Fig. 1, under the Assumptions 1.
U.t.c.

Σ_2 is exp. stable

iff

the char. poly. χ of Σ_2, given by (26), satisfies

33 $$Z[\chi] \subset \overset{\circ}{\mathbb{C}}_- = \mathbb{C} \backslash \mathbb{C}_+ .$$

or, equiv. iff

Σ_2 has no eigenvalues λ in $\mathbb{C}_+$. ■

34 <u>Comment</u>. Note that $\mathbb{C}_+$, i.e., the <u>closed</u> right-half of the complex plane, is the undesired set of the complex plane, where Σ_2 must not have any eigenvalue for Σ_2 to be exp. stable. Now for technical reasons (e.g., speed of response, damping, etc.) it is desirable to have a larger undesirable set, say $\mathcal{U} \subset \mathbb{C}$ as, for example, in Fig. 2.

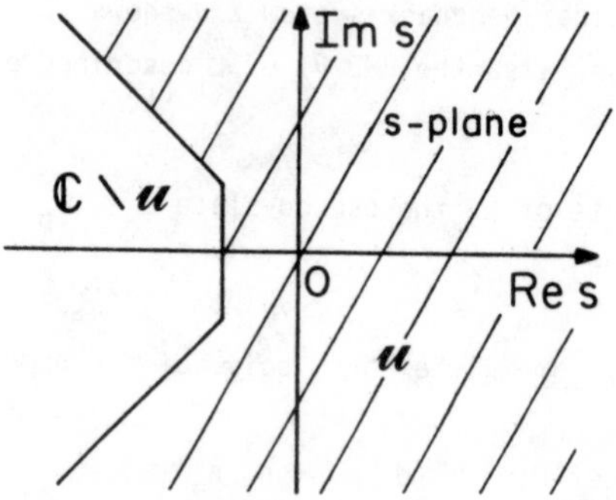

Fig. 2. $\mathbb{C}$ divided in an undesirable subset $\mathcal{U}$ and a desirable subset $\mathbb{C} \backslash \mathcal{U}$.

Hence the following stronger requirement for stability.

36 Definition. Consider feedback system Σ_2 shown in Fig. 1. Let $\mathcal{U}$ be a closed subset of $\mathbb{C}$ which is symmetric w.r.t. the real axis and such that $\mathcal{U} \supset \mathbb{C}_+$. We say that Σ_2 is $\mathcal{U}$-stable iff its PMD $\mathcal{D}_g$, described by (11)-(18), is well formed and its closed-loop eigenvalues are in $\mathbb{C}\backslash\mathcal{U}$.

In view of Fact 25 and Definitions 29, 30, and 36, we now have

37 Theorem [$\mathcal{U}$-stability of Σ_2]. If the Assumptions 1 hold, then

feedback system Σ_2, shown in Fig. 1, is $\mathcal{U}$-stable

iff

the char. poly. χ, given by (26), satisfies

38 $$Z[\chi] \subset \mathbb{C}\backslash\mathcal{U}.$$ ∎

Let us now consider the input-output properties of Σ_2.

Let

39 $$\mathcal{R}_{\mathcal{U}} := \{f \in \mathbb{R}_p(s) : f \text{ is analytic in } \mathcal{U}\} \subset \mathcal{R}(0)$$

denote the subring of exp. stable transfer functions (2.2.6), that are analytic in $\mathcal{U}$.

40 Corollary. Under the assumptions and condition of Theorem 37, feedback system Σ_2 has the property that all its closed-loop transfer functions have elements in $\mathcal{R}_{\mathcal{U}}$, more precisely the closed-loop transfer function

$$H : \begin{bmatrix} u_1 \\ u_2 \\ v_1 \\ d_0 \end{bmatrix} \mapsto \begin{bmatrix} y_1 \\ y_2 \\ e_1 \\ e_2 \end{bmatrix} \in \mathcal{R}_{\mathcal{U}}^{4\times4}.$$

41 Exercise. Prove Corollary 40.

42 Exercise. Consider the feedback system Σ_2. Let Assumptions 1 hold and let Σ_2 be $\mathcal{U}$-stable. Show that the $\mathcal{U}$-stability of Σ_2 still holds when the coefficients of the polynomials n_π, n_f, n_p, d_p, and d_c are subjected to

sufficiently small perturbations. (Hint: Note that $\mathcal{U}$ is closed.)

We shall now study conditions which ensure that the compensator $c = [\pi \mid f]$ is proper. Consider Fig. 1; let q denote closed-loop transfer function $q : u_1 \mapsto y_1$ and let $h_{y_2 v_1}$ denote the system I/O map $h_{y_2 v_1} : v_1 \mapsto y_2$. Straightforward calculations show that

45 $$q = f(1 + pf)^{-1} = d_p \chi^{-1} n_f$$

and

46 $$h_{y_2 v_1} = p(1 + fp)^{-1}\pi = n_p\, \chi^{-1}\, n_\pi .$$

47 Theorem [Properness of f and π]. Consider system Σ_2 shown in Fig. 1. Let $p \in \mathbb{R}_{p,o}(s)$.
U.t.c.

(a)

48 $f \in \mathbb{R}_p(s)$ $(\mathbb{R}_{p,o}(s)$, resp.);

iff

49 $q \in \mathbb{R}_p(s)$ $(\mathbb{R}_{p,o}(s)$, resp.).

(b) Let $f \in \mathbb{R}_p(s)$; then

50 $\pi \in \mathbb{R}_p(s)$ $(\mathbb{R}_{p,o}(s)$, resp.);

iff

51 $p^{-1} h_{y_2 v_1} \in \mathbb{R}_p(s)$ $(\mathbb{R}_{p,o}(s)$, resp.).

52 Comment. In view of (46), implication (50) ⇒ (51) is intuitively obvious: it says that if the compensator $c = [\pi \mid f]$ is proper, then the I/O map $h_{y_2 v_1} \to 0$ as $|s| \to \infty$ at least as fast as the plant p. Furthermore, this is a necessary requirement on $h_{y_2 v_1}$ in order to obtain a proper compensator.

53 Proof of Theorem 47

(a) ⇒ : Since p is strictly proper, if f is proper, then q is proper by (45).

(a) ⇐ : Equation (45) is equivalent to

54 $$f = q(1 - pq)^{-1};$$

hence, since p is strictly proper, q proper implies that f is proper.

(b) ⇒ : f proper and π proper imply that

$$(1 + fp)^{-1}\pi \in \mathbb{R}_p(s);$$

hence (51) follows from (46).

(b) ⇐ : By assumption p is strictly proper, f is proper, and (46) can be written as

55 $$(1 + fp)\, p^{-1} h_{y_2v_1} = \pi;$$

hence (51) ⇒ π is proper. ∎

From Corollary 40 and Theorem 47 we can state properties of the achievable I/O maps:

56 Corollary. Consider feedback system Σ_2, shown in Fig. 1, under the Assumptions 1 and let Σ_2 be $\mathcal{U}$-stable. Then

46 $$h_{y_2v_1} = p(1 + fp)^{-1}\pi = n_p\chi^{-1}n_\pi$$

satisfies:

57 $$p^{-1}h_{y_2v_1} \in \mathbb{R}_p(s),$$

58 $$\mathcal{P}[h_{y_2v_1}] \subset \mathbb{C}\backslash\mathcal{U},$$

59 $$\mathcal{Z}[h_{y_2v_1}] \cap \mathcal{U} = \{\mathcal{Z}[n_p] \cup \mathcal{Z}[n_\pi]\} \cap \mathcal{U}.$$

In other words, any $\mathcal{U}$-zero of p and any $\mathcal{U}$-zero of π is necessarily a $\mathcal{U}$-zero of $h_{y_2v_1}$ because in (46) χ cannot cancel any $\mathcal{U}$-zero of p or π.

60 Exercise. For the system Σ_2, let Assumptions 1 hold and let Σ_2 be $\mathcal{U}$-stable. Prove the following:

(a) Let $H_{yu} : (u_1, u_2, v_1) \mapsto (y_1, y_2)$, then, with q defined by (45),

$$H_{yu} = \left[\begin{array}{c|c|c} f(1+pf)^{-1} & -fp(1+fp)^{-1} & (1+fp)^{-1}\pi \\ \hline pf(1+pf)^{-1} & p(1+fp)^{-1} & p(1+fp)^{-1}\pi \end{array}\right],$$

$$= \left[\begin{array}{c|c|c} q & -qp & (1-qp)\pi \\ \hline pq & p(1-qp) & p(1-qp)\pi \end{array}\right].$$

(b) If $H_{yu} \in \mathcal{R}_{\mathcal{U}}^{2\times 3}$, then all transfer functions from any (u_1, u_2, v_1, d_0) to any $(y_1, y_2, e_1, e_2, y_p)$ are in $\mathcal{R}_{\mathcal{U}}$.

(c) Assume, in addition, that $p \in \mathcal{R}_{\mathcal{U}} \cap \mathbb{R}_{p,o}(s)$; then the I/O map $h_{y_2v_1}$ can be realized by a $\mathcal{U}$-stable feedback system Σ_2 with π and $f \in \mathbb{R}_p(s)$ if and only if

$$h_{y_2v_1} = p(1-qp)\pi$$

for some $q \in \mathcal{R}_{\mathcal{U}}$ and $\pi \in \mathbb{R}_p(s)$ s.t. $(1-qp)\pi \in \mathcal{R}_{\mathcal{U}}$. ✖

We shall now study robust asymptotic tracking throughout, the inputs v_1 to be tracked (specified in terms of Laplace transforms) are in the class

61 $$\Psi := \{\hat{v}_1 = \psi^{-1}\mu : \psi, \mu \in \mathbb{R}[s] \text{ with } \partial[\mu] < \partial[\psi]\},$$

where ψ is a given monic polynomial and μ is arbitrary; furthermore,

62 $$Z[\psi] \subset \mathbb{C}_+ \subset \mathcal{U}$$

and

63 $$Z[\psi] \cap Z[n_p] = \phi.$$

The arbitrariness of μ in (61) corresponds to selecting arbitrary initial conditions in the dynamical system generating all the reference inputs v_1 in the class Ψ.

We will see later that (63) is a necessary and sufficient condition that the plant $p \in \mathbb{R}_{p,o}(s)$ must satisfy in order that the system of Fig. 1 with a proper compensator $c = [\pi \,\vdots\, f]$ can track every input in the class Ψ. (This is a well-known fact for the multivariable case; see, e.g. [Dav.1], [Des.2].)

Let

64[a] $$\eta(t) := -(y_2 - v_1)(t)$$

be the tracking error (see Fig. 1), whence in terms of the Laplace transform

64[b] $$\hat{\eta} = -(h_{y_2v_1} - 1)\hat{v}_1 .$$

65 Definition. We say that feedback system Σ_2, shown in Fig. 1, achieves robust asymptotic tracking over the class Ψ if and only if the following three conditions are satisfied:

(1) system Σ_2, shown in Fig. 1, is $\mathcal{U}$-stable;

66 (2) for any input $v_1 \in \Psi$, the tracking error $\eta(t) \to 0$ exponentially as $t \to \infty$;

(3) the tracking requirement (66) holds for any perturbed plant $\tilde{p} = \tilde{n}_p \tilde{d}_p^{-1} \in \mathbb{R}_{p,o}(s)$ where the polynomials $\tilde{n}_p$ and $\tilde{d}_p$ are arbitrary subject to

(i) $\tilde{n}_p$ and $\tilde{d}_p$ are coprime;

(ii) $Z[\psi] \cap Z[\tilde{n}_p] = \phi$;

(iii) the perturbed closed-loop system Σ_2 is exp. stable (i.e. $Z[\tilde{\chi}] \subset \mathring{\mathbb{C}}_-$ where $\tilde{\chi} := \tilde{d}_p d_c + \tilde{n}_p n_f$). ∎

Now, for any input $v_1 \in \Psi$, (64[b]) and (46) give

70 $$\hat{\eta} = \frac{[n_p(n_\pi - n_f) - d_p d_c]}{\chi} \frac{\mu}{\psi}$$

From (70), we obtain the last theorem of this section.

71 Theorem [Robust asymptotic tracking]. Consider feedback system Σ_2, shown in Fig. 1, under the Assumptions 1. Let the class Ψ of inputs v_1 to be tracked be specified by (61), (62), (63).

U.t.c.

72 feedback system Σ_2 achieves robust asymptotic tracking over the class Ψ

iff

73 (i) $Z[\chi] \subset \mathbb{C}\backslash\mathcal{U}$;

74 (ii) $\psi | d_c$;

75 (iii) $\psi | (n_\pi - n_f)$.

76 Comments. (a) Since $\chi = d_p d_c + n_p n_f$, (73) and (74) imply that

63 $$Z[\psi] \cap Z[n_p] = \phi.$$

Algorithm (5.3.55) below shows the converse; namely, if (63) holds, then there exists a proper f and π such that robust asymptotic tracking over the class Ψ is achieved .

(b) From (70), conditions (73), (74), and (75) imply that $\forall\zeta \in Z[\psi]$, $h_{y_2v_1}(\zeta) = 1$; and, if ζ is a kth order zero of ψ, then $h^{(i)}_{y_2v_1}(\zeta) = 0$ for $i = 1, 2, \cdots, k - 1$. This is as expected from Exercise 3.3.2.28.

78 Proof of Theorem 71. $\Leftarrow$. From (73)-(75) and Corollary 40, it follows that $\hat{n}$, given by (70), is strictly proper and analytic in $\mathbb{C}_+$; hence (66) follows. Furthermore, this conclusion still holds for any $\tilde{n}_p$, $\tilde{d}_p$ as long as $Z[\tilde{\chi}] \subset \mathring{\mathbb{C}}_-$, as required by the definition of robust asymptotic tracking.

$\Rightarrow$. Robust asymptotic tracking implies (73)-(75): indeed, (73) is required for $\mathcal{U}$-stability (Theorem 37), and the robust asymptotic tracking condition (66), for any arbitrary $\tilde{n}_p$, $\tilde{d}_p$ as in Definition 65, requires ψ to cancel in (70) for any such $\tilde{n}_p$, $\tilde{d}_p$. Hence (74)-(75) must hold, i.e., $\psi | d_c$ and $\psi | n_\pi - n_f$

More precisely, consider the case where ψ has a simple zero at $\zeta \in \mathbb{C}_+$. Since robust asymptotic tracking is achieved, from (70) we have

79 $$n_p(\zeta)\big(n_\pi(\zeta) - n_f(\zeta)\big) - d_p(\zeta)d_c(\zeta) = 0$$

and (79) holds when arbitrary small perturbations are applied to the coefficients of n_p and d_p. Consider two cases:

Case 1. $d_p(\zeta) \neq 0$. Consider the following perturbation in n_p and d_p: for some sufficiently small $\varepsilon > 0$, multiply all the coefficients of d_p by $(1 - \varepsilon)$; then, by robust asymptotic tracking, we have

80 $$n_p(\zeta)\big(n_\pi(\zeta) - n_f(\zeta)\big) - (1 - \varepsilon)d_p(\zeta)d_c(\zeta) = 0.$$

By assumption $d_p(\zeta) \neq 0$, and, by (63), $n_p(\zeta) \neq 0$; hence equations (79) and (80) imply that

81 $$n_\pi(\zeta) - n_f(\zeta) = d_c(\zeta) = 0.$$

Case 2. $d_p(\zeta) = 0$. By assumption, $\partial[d_p] \geq 1$, so an arbitrarily small perturbation of a suitable coefficient of d_p yields a new polynomial $\tilde{d}_p$ such that $\tilde{d}_p(\zeta) \neq 0$; robust asymptotic tracking then implies that

82 $$n_p(\zeta)\big(n_\pi(\zeta) - n_f(\zeta)\big) - \tilde{d}_p(\zeta)d_c(\zeta) = 0.$$

Since $n_p(\zeta) \neq 0$, $\tilde{d}_p(\zeta) \neq 0$, and $d_p(\zeta) = 0$, (79) and (82) imply (81).

Therefore, we have proven that robust asymptotic tracking implies (74) and (75), when ψ has simple zeros. When ψ has multiple zeros, a similar proof can be constructed. ∎

5.3. Design

We propose to use the theorems of Sec. 5.2 to show how one may obtain a compensator for system Σ_2, shown in Fig. 5.2.1, to achieve a prescribed I/O map $h_{y_2v_1}$ and to achieve robust asymptotic tracking over a given class Ψ.

8 Data. We are given

9 (1) the closed set $\mathcal{U} \supset \mathbb{C}_+$ of undesirable closed-loop eigenvalue locations

(2) the plant

10 $$p = \frac{n_p}{d_p} = \frac{n_{pu}\, n_{ps}}{d_p} \in \mathbb{R}_{p,o}(s),$$

where

11 (i) n_p and d_p are coprime,

12 (ii) $n_p = n_{pu}\, n_{ps}$ with $Z[n_{pu}] \subset \mathcal{U}$, $Z[n_{ps}] \subset \mathbb{C}\backslash\mathcal{U}$;

(3) the specified I/O map $h_{y_2v_1}$ is given by

13 $$h_{y_2v_1} = \frac{n_{pu}\, n_{h1}}{d_h}$$

where

14 (i) $p^{-1}h_{y_2v_1} \in \mathbb{R}_p(s)$ (so that by Theorem 5.2.47 the compensator will be proper);

15 (ii) (n_{h_1}, d_h) are coprime polynomials;

16 (iii) $Z[d_h] \subset \mathbb{C}\backslash\mathcal{U}$ (a consequence of the required $\mathcal{U}$-stability of the system).

20 Comments. (a) Recall that, from (5.2.26), $\chi = d_c d_p + n_f n_p$ and that by (5.2.45) and (5.2.46),

21 $$q = d_p\,\chi^{-1}\,n_f,$$

22 $$h_{y_2v_1} = n_p\,\chi^{-1}\,n_\pi.$$

(b) By Theorem 5.2.47 and (14) once q in (21) is proper, the compensator $c = [\pi \mid f]$ is proper, and then by Theorem 5.2.37 and (10), the $\mathcal{U}$-stability of feedback system Σ_2 is equivalent to $Z[\chi] \subset \mathbb{C}\backslash\mathcal{U}$. Therefore, in (22), any factor common to n_p and χ must have all its zeros in $\mathbb{C}\backslash\mathcal{U}$. Hence n_{pu} must be a factor of the numerator of $h_{y_2v_1}$ as indicated in (13).

(c) Equating the two expressions (13) and (22) for $h_{y_2v_1}$, and canceling n_{pu}, we get

23 $$n_\pi = \frac{n_{h_1}\,\chi}{d_h\,n_{ps}}\;.$$

Note that the $h_{y_2v_1}$ specified by (13) will be achieved by the closed-loop system of Fig. 1 if and only if (23) holds.

(d) The algorithm below actually establishes the converse of Corollary 5.2.56: indeed, conditions (13)-(16) are identical with (5.2.46), (5.2.57)-(5.2.59).

24 Algorithm [For Σ_2 to achieve the prescribed I/O map $h_{y_2v_1}$].

Step 1. Choose $\chi \in \mathbb{R}[s]$ s.t.

25 (i) $d_h\,n_{ps} \mid n_{h_1}\chi$ (hence (23) will give n_π as a polynomial);

26 (ii) $Z[\chi] \subset \mathbb{C}\backslash\mathcal{U}$ (to achieve $\mathcal{U}$-stability);

27 (iii) $\partial\chi \geq 2\,\partial d_p - 1$. (See Comment (35)(b), below.)

Step 2. Choose $n_f \in \mathbb{R}[s]$ s.t.

30 (i) $\partial n_f \leq \partial\chi - \partial d_p$ (equiv. q, in (21), is proper);

31 (ii) $d_p | (\chi - n_p n_f)$

and set

32 $$d_c := \frac{\chi - n_p n_f}{d_p}$$

33 $$n_\pi := \frac{n_{h_1}\,\chi}{d_h\, n_{ps}}\,.$$

End of Algo

35 Comments. (a) The first step is to choose the closed-loop characteristic polynomial. Note that the algorithm does not guarantee that n_f and d_c are coprime, nor that $[n_\pi \,\vdots\, n_f]$ and d_c are coprime. Examples show that complete cancellation may occur! Note that in any case, the analysis in (5.2.10) and the subsequent Definition 5.2.29 establish that $d_c d_p + n_f n_p$ is the characteristic polynomial of (the PMD $\mathcal{D}_g$ of) Σ_2. Of course, cancellation of common factors in $[n_\pi \,\vdots\, n_f]$ and d_c will simplify the realization of the compensator and the resulting closed-loop system will still have all its closed-loop eigenvalues in $\mathbb{C}\backslash\mathcal{U}$ (see (26)).
(b) Note that (31)-(32) is equivalent to the solution of the equation

36 $$n_f n_p + d_c d_p = \chi,$$

where n_p, d_p, and χ are given and n_f, d_c are unknown. Since (n_p, d_p) is coprime, the Bezout identity states that there are polynomials u, v (obtained by the Euclidean algorithm [MacL.1, p. 122]) such that

37 $$u n_p + v d_p = 1.$$

Hence, all polynomial solutions (n_f, d_c) of (36) are given by

38 $$\begin{aligned} n_f &= u\chi - d_p m \\ d_c &= v\chi + n_p m \end{aligned} \qquad \text{for any } m \in \mathbb{R}[s].$$

(To see this: multiply (37) by χ to obtain a particular solution $(u\chi, v\chi)$ of (36) and add to it the general solution $(-d_p m, n_p m)$ of the homogeneous equation corresponding to (36); see also Theorem 6.2.39 below.) Conditions under which (38) yields a proper feedback compensator $f = n_f/d_c$ are easily obtained: by (21) and Theorem 5.2.47 we have

30 $$\partial[n_f] \leq \partial[\chi] - \partial[d_p]$$

$\Leftrightarrow$ q proper $\Leftrightarrow$ f proper.

Hence by Theorem 5.2.47, (14) and (30) are necessary and sufficient conditions in order that the compensator $c = [\pi \,\vdots\, f]$ be proper. Moreover, under these conditions and since $p \in \mathbb{R}_{p,o}(s)$ it follows from (36) that

39 $$\partial[d_c] = \partial[\chi] - \partial[d_p].$$

Observe now that the least degree solution n_f of (36) is given by (38), by dividing $u\chi$ by d_p (note that m and n_f are then resp. quotient and remainder). Hence for that solution

40 $$\partial[n_f] \leq \partial[d_p] - 1.$$

So if we apply condition (27), equiv.

41 $$\partial[d_p] - 1 \leq \partial[\chi] - \partial[d_p],$$

then we have that the least degree solution n_f satisfies (30). Hence (27) is a sufficient condition for the existence of a proper feedback compensator f characterized by (38) and (30), or equiv. (30)-(32), and the precompensator π is then proper by (14) and Theorem 5.2.47.

(c) In (31) χ and n_p are fixed; hence we have to adjust the coefficients of $n_f(s) := \sum_{j=0}^{n} n_{f,j}s^j$ so that $\chi - n_p\, n_f$ is a multiple of d_p.

For simplicity suppose that d_p has $m := \partial[d_p]$ distinct zeros $(p_k)_{k=1}^m$; then (31) is equivalent to

42 $$\chi(p_k) - n_p(p_k)n_f(p_k) = 0 \quad \forall k = 1, 2, \cdots, m = \partial[d_p].$$

Since χ and n_p are known polynomials, with $n_p(p_k) \neq 0$ $\forall k$ (because (n_p, d_p) is coprime), (31) $\Leftrightarrow$ (42) is a system of m linear equations in n + 1 unknowns $(n_{f,j})_0^n \in \mathbb{R}^{n+1}$; moreover, if $n \geq m - 1$, the system (42) has a matrix with a nonsingular $m \times m$ Vandermonde submatrix, (since $p_k \neq p_\ell$, $\forall k \neq \ell$). Hence for $n \geq m - 1$, (42) has (a) solution(s) $(n_{f,j})_0^n$ and n_f can be computed in this manner; moreover, the solution (40) is obtained if we set $n = m - 1 = \partial[d_p] - 1$.

Summary. Algorithm 24 delivers the specified I/O map $h_{y_2v_1}$: the design gives a proper $f = n_f/d_c$, a proper $\pi = n_\pi/d_c$, and a $\mathcal{U}$-stable char. poly. χ.

45 Example. Given $p(s) = (s + 1)/(s - 1)^3$. To achieve the I/O map $h_{y_2v_1} = (s + 1)^{-2}$, we apply Algorithm 24: let $\chi(s) = (s + 1)^5$; then $n_\pi(s) = (s + 1)^2$, $n_f(s) = 8(3s^2 - 2s + 1)$ and $d_c(s) = (s + 1)(s + 7)$.

The design of compensators for robust asymptotic tracking is based on Theorem 5.2.71. It is best explained in terms of an algorithm-

55 Algorithm [Robust asymptotic tracking]

Data:

56 (1) $\mathcal{U}$ and p are specified as in (9)-(12),

57 (2) $\psi \in \mathbb{R}[s]$ with $Z[\psi] \subset \mathbb{C}_+$ and

58 $$Z[\psi] \cap Z[n_p] = \phi.$$

Step 1. Choose $\chi \in \mathbb{R}[s]$ s.t.

61 $Z[\chi] \subset \mathbb{C}\backslash\mathcal{U}$,

62 $\partial[\chi] \geq \partial\psi + 2\partial d_p - 1$

Step 2. Choose $n_f \in \mathbb{R}[s]$ s.t.

65 $\partial n_f \leq \partial\chi - \partial d_p$,

66 $(\psi d_p) \mid (\chi - n_f n_p)$.

Set

67 $$d_c := \frac{\chi - n_f n_p}{d_p} .$$

<u>Step 3</u>. Choose $n_\pi \in \mathbb{R}[s]$ s.t.

70 $\partial n_\pi \leq \partial\chi - \partial d_p$,

71 $\psi | (n_\pi - n_f)$.

<u>Justification</u>

By (67), $\chi = d_c d_p + n_f n_p$ is the char. poly. of the closed-loop system. Robust asymptotic tracking is achieved: indeed, recall Theorem 5.2.71: note (61); by (66) and (67), $\psi | d_c$; by (71) $\psi | (n_\pi - n_f)$.

Now recall Theorem 5.2.47: in view of (21), (65) implies that q is proper, hence f is proper. Hence by (67) $\partial[d_c] = \partial[\chi] - \partial[d_p]$, s.t. (70) implies that π is proper.

Note that by (62) and (65), $n_f = \sum_{j=0}^{n} n_{f,j} s^j$ has a coefficient vector $(n_{f,j})_0^n$ s.t. n may be chosen s.t. $n = \partial\psi + \partial d_p - 1$; therefore, there are enough parameters to satisfy (66). ∎

Chapter 6. The Closed-Loop Eigenvalue Placement Problem

6.1. Introduction

In this chapter we are given a strictly proper multi-input multi-output plant P and we wish to design a unity feedback system Σ using a compensator C. This compensator C should achieve a given closed-loop eigenvalue placement, which is specified in terms of a nonsingular polynomial matrix D_k resulting in the closed-loop characteristic polynomial $\chi(s) = \det D_k(s)$. D_k specifies completely the zero-input pseudo-state trajectories of feedback system Σ (Fact 6.2.30). The compensator, being a polynomial matrix fraction, has a numerator and denominator that satisfy compensator equation (6.2.32). Theorem 6.2.39 describes the class of polynomial matrix solutions of this equation. Theorem 6.2.61 characterizes all internally proper solutions. Finally, Theorem 6.2.84 gives sufficient conditions for the existence of internally proper solutions.

6.2. The Compensator Problem

1 Problem COMP. We are given a plant P s.t.

2 $P \in \mathbb{R}_{p,o}(s)^{n_o \times n_i}$ has an int. pr. r.c.f. (N_{pr}, D_{pr}) s.t.

3 D_{pr} is column-reduced with column degrees k_j, $j \in \underline{n}_i$,
and
4 with highest column degree coefficient matrix $D_{ph} = I_{n_i}$.

We must find a compensator C s.t.

5 $C \in \mathbb{R}_p(s)^{n_i \times n_o}$ has an int. pr. ℓ.f. $(D_{c\ell}, N_{c\ell})$
with

6 $D_{c\ell}$ row-reduced,

s.t.

7 the feedback system Σ, shown in Fig. 4.3.1, is exp. stable with prescribed closed-loop characteristic polynomial $\chi(s)$.

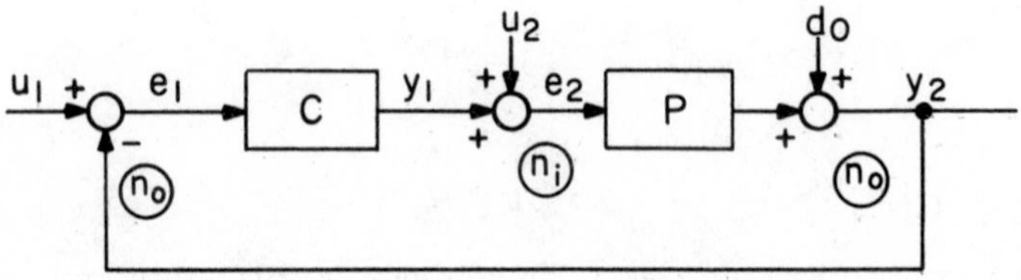

Fig. 4.3.1. The feedback system Σ under consideration.

8 Comments. (a) An int. pr. r.c.f. of P as in (2), (3) can always be obtained from any r.f. of P by using suitable e.c.o.'s to extract a g.c.r.d. and make D_{pr} column-reduced.

(b) Assumption (3) is equivalent to

9 $$D_{pr} = D_{p-} \operatorname{diag}[s^{k_j}]_{j=1}^{n_i},$$

where D_{p-} is biproper and

10 $$D_{p-}(\infty) = D_{ph}.$$

(c) In order to achieve $D_{ph} = I$ as in Assumption (4), a change of plant-input-coordinates may be necessary. The plant gives

11 $$y_2 = Pe_2 = N_{pr} D_{pr}^{-1} e_2.$$

If $D_{ph} \neq I$, we set

12 $$\bar{e}_2 = D_{ph}^{-1} e_2;$$

hence

13 $$y_2 = \bar{P}\,\bar{e}_2, \quad \text{where} \quad \bar{P} = N_{pr}(D_{ph}^{-1} D_{pr})^{-1}.$$

(d) By Corollary 4.3.26, the feedback Σ obeys assumptions IS, WP, and WF of Sec. 4.2; furthermore, Σ is exp. stable if and only if $Z[\chi] \subset \mathring{\mathbb{C}}_-$, where $\chi(s)$ equals the RHS of (4.3.8). ∎

16 Analysis. Analyzing Σ, as in Sec. 4.3, we obtain

17 $$\chi(s) = \det[D_{c\ell}D_{pr} + N_{c\ell}N_{pr}](s),$$

18 $$H_{y_2u_1} = N_{pr}[D_{c\ell}D_{pr} + N_{c\ell}N_{pr}]^{-1}N_{c\ell}.$$

Thus setting

19 $$D_k := D_{c\ell}D_{pr} + N_{c\ell}N_{pr},$$

we have

20 $$\chi(s) = \det D_k(s).$$

Let us view P and C as specified by the PMD's $\mathcal{D}_p = [D_{pr}, I, N_{pr}, 0]$ and $\mathcal{D}_c = [D_{c\ell}, N_{c\ell}, I, 0]$, resp.. Call ξ_p, ξ_c, resp., the corresponding pseudo-states. As in Secs. 4.2 and 4.3, we consider the PMD $\mathcal{D}_{yu} = [D_g, N_\ell, N_r, 0]$ where $\xi := (\xi_c^T, \xi_p^T)^T$ and

25 $$N_r = \text{diag}[I \mid N_{pr}],$$

26 $$N_\ell = \text{diag}[N_{c\ell} \mid I],$$

27 $$D_g = \left[\begin{array}{c|c} D_{c\ell} & N_{c\ell}N_{pr} \\ \hline -I & D_{pr} \end{array}\right].$$

For zero input, the differential equation representing the system reads (after performing on (27) the e.r.o. $\rho_1 \leftarrow \rho_1 + D_{c\ell}\rho_2$),

28 $$\begin{bmatrix} 0 & D_k \\ -I & D_{pr} \end{bmatrix}\begin{bmatrix} \xi_c(t) \\ \xi_p(t) \end{bmatrix} = \begin{bmatrix} \theta \\ \theta \end{bmatrix}$$

or equivalently,

29 $$\begin{cases} D_k(p)\xi_p(t) = \theta, \\ \xi_c(t) = D_{pr}(p)\xi_p(t). \end{cases}$$

From (20), (25), (26), and (29), we deduce the following results:

30 Fact. The polynomial matrix $D_k = D_{c\ell}D_{pr} + N_{c\ell}N_{pr}$ specifies completely the z-i p.s trajectories $\xi(\cdot)$ of the feedback system Σ. By (29), D_k controls directly the dynamics of the z-i p.s. trajectories $\xi_p(\cdot)$ of the plant. By (20) the determinant of D_k is the char. pol. of Σ. ∎

31 Comment. If D_k is chosen to be diagonal, then each diagonal entry specifies independently the characteristic polynomial of each entry of $\xi_p(\cdot)$

Solving for the Compensator. In design, N_{pr} and D_{pr} are given and they satisfy (1)-(4). The matrix D_k is chosen so that $Z[\det D_k(s)] \subset \mathring{\mathbb{C}}_-$, for exp. stability, and so that the compensator equation

32 $$X\,D_{pr} + Y\,N_{pr} = D_k$$

has (a) solution(s)

33 $$(X, Y) \in \mathbb{R}[s]^{n_i \times n_i} \times \mathbb{R}[s]^{n_i \times n_o},$$

s.t.

34 the ℓ.f. (X, Y) is int. pr. and X is row reduced. ∎

Note now that, since by (2)-(4), (N_{pr}, D_{pr}) is a r.c.f. of P, Theorem 2.4.1.25.R delivers a generalized Bezout identity: there exist six polynomial matrices

35 $$U_{pr}, V_{pr}, D_{p\ell}, N_{p\ell}, U_{p\ell}, V_{p\ell} \in E(\mathbb{R}[s]),$$

s.t.

36 $$\begin{array}{c} \\ n_i \\ n_o \end{array}\!\!\begin{array}{c} \begin{array}{cc} n_i & n_o \end{array} \\ \left[\begin{array}{c|c} V_{pr} & U_{pr} \\ \hline -N_{p\ell} & D_{p\ell} \end{array}\right] \end{array} \begin{array}{c} \begin{array}{cc} n_i & n_o \end{array} \\ \left[\begin{array}{c|c} D_{pr} & -U_{p\ell} \\ \hline N_{pr} & V_{p\ell} \end{array}\right] \end{array} = \left[\begin{array}{c|c} I_{n_i} & 0 \\ \hline 0 & I_{n_o} \end{array}\right].$$

Moreover,

37 $$(D_{p\ell}, N_{p\ell}) \text{ is a } \ell\text{.c.f. of } P$$

and without loss of generality

38 $$D_{p\ell} \text{ is row-reduced.}$$

39 Theorem [Polynomial solutions of comp. eq. (32)]. Consider (32), where $P = N_{pr}D_{pr}^{-1}$ satisfies (2)-(4). U.t.c.

40 $$(X, Y) \in \mathbb{R}[s]^{n_i \times n_i} \times \mathbb{R}[s]^{n_i \times n_o} \text{ is a solution of (32)}$$

$\Leftrightarrow$

$$\exists\, N_k \in \mathbb{R}[s]^{n_i \times n_i} \quad \text{s.t.}$$

41 $$X = D_k V_{pr} - N_k N_{p\ell},$$

42 $$Y = D_k U_{pr} + N_k D_{p\ell},$$

where U_{pr}, V_{pr}, $D_{p\ell}$, $N_{p\ell}$ are elements of the generalized Bezout identity (36). Moreover,

43 $$(X, Y) \text{ is } \ell.\text{c.} \Leftrightarrow (D_k, N_k) \text{ is } \ell.\text{c.}$$ ∎

44 Comments. (a) Given a solution (X, Y), we set $(X, Y) = (D_{c\ell}, N_{c\ell})$, i.e., we obtain the compensator $C = D_{c\ell}^{-1}N_{c\ell} = X^{-1}Y$, provided that $\det X \neq 0$. The additional requirement that (X, Y) be an int. pr. ℓ.f. with X row-reduced is investigated below.

(b) Equations (41) and (42) constitute a global parametrization by the polynomial matrix N_k of all the solutions of (32): N_k determines (X, Y) uniquely, and vice versa. Note that N_k may be viewed as the quotient of the division of $D_k U_{pr}$ on the right by $D_{p\ell}$: Write (42) as

45 $$D_k U_{pr} = -N_k D_{p\ell} + Y \quad \text{s.t.} \quad YD_{p\ell}^{-1} \in \mathbb{R}_{p,o}(s)^{n_i \times n_o}.$$

The quotient N_k then determines X by (41).

46 Proof of Theorem 39. (a) Proof of $\Leftarrow$: By assumption (41) and (42) hold. For any N_k substitute X and Y given by (41) and (42) into (32); for the LHS of (32) we obtain

47 $$D_k(V_{pr}D_{pr} + U_{pr}N_{pr}) + N_k(-N_{p\ell}D_{pr} + D_{p\ell}N_{pr}).$$

Now considering the (1, 1) entry and the (2, 1) entry of (36), (47) reduces to D_k, as required by (32). Hence $\forall N_k$, X and Y, defined by (41) and (42), are solutions of (32).

(b) Proof of $\Rightarrow$: By assumption (X, Y) is a solution of (32). First, consider the (1, 1) entry of (36) and multiply on the left by D_k; then

$$D_k V_{pr} D_{pr} + D_k U_{pr} N_{pr} = D_k.$$

Hence

48 $$X_p := D_k V_{pr}, \; Y_p := D_k U_{pr}$$

specifies a particular solution of (32).

Second, since (32) is a linear equation we seek the general solution of the homogeneous equation

49 $$X\, D_{pr} + Y\, N_{pr} = 0.$$

Recall that the (2, 1) entry of (36) gives $-N_{p\ell}D_{pr} + D_{p\ell}N_{pr} = 0$. For any polynomial solution (X_h, Y_h) of the homogeneous equation (49) let

50 $$N_k := Y_h D_{p\ell}^{-1};$$

then, since (X_h, Y_h) is a solution of (49),

$$X_h = -Y_h\, N_{pr}\, D_{pr}^{-1} = -Y_h\, D_{p\ell}^{-1}\, N_{p\ell} = -N_k\, N_{p\ell}.$$

Hence any solution (X_h, Y_h) of (49) is of the form

51 $$X_h = -N_k\, N_{p\ell}, \qquad Y_h = N_k\, D_{p\ell}.$$

Finally, we claim that N_k, defined by (50), is a polynomial matrix; indeed, using (36),

$$N_k = N_k(N_{p\ell}U_{p\ell} + D_{p\ell}V_{p\ell}) = -X_h U_{p\ell} + Y_h V_{p\ell} \in E(\mathbb{R}[s]).$$

Hence we have exhibited a particular solution (X_p, Y_p) in (48) and we have shown that <u>any</u> solution (X_h, Y_h) is of the form (51), where N_k is a <u>polynomial</u> matrix; consequently any solution (X, Y) of (32) is of the form (41) and (42).

(c) From (41) and (42) we have

52 $$[X \mid Y] = [D_k \mid N_k]\begin{bmatrix} V_{pr} & U_{pr} \\ -N_{p\ell} & D_{p\ell} \end{bmatrix}.$$

The last matrix in the RHS is unimodular (see (36)); hence $\mathrm{rk}[X \mid Y](s) = \mathrm{rk}[D_k \mid N_k](s)$, $\forall s \in \mathbb{C}$ and (43) follows. ∎

53 <u>Exercise.</u> Prove the following. If in the analysis (16) we had used the characteristic polynomial (4.3.9), we would have obtained the comp. eqn. $D_{p\ell}X + N_{p\ell}Y = D_k$. All polynomial solutions of this equation are of the form

$$X_r = V_{p\ell}D_k - N_{pr}N_k$$

$$Y_r = U_{p\ell}D_k + D_{pr}N_k.$$ ∎

We now give conditions under which the solutions (X, Y) of the comp. equ. (32) are s.t. $X^{-1}Y$ is an int. proper left fraction, [Emr.1].

61 <u>Theorem</u> [Internally proper solutions of comp. equ. (32)]. Consider the comp. equ. (32), where the plant P is given by

2 $P \in \mathbb{R}_{p,o}(s)^{n_o \times n_i}$ has an int. pr. r.c.f. (N_{pr}, D_{pr})

s.t.

3 D_{pr} is column reduced with column degrees k_j, $j \in \underline{n}_i$

and

4 with highest column degree coefficients matrix $D_{ph} = I_{n_i}$.

U.t.c.

the compensator equation, given by

32 $$XD_{pr} + YN_{pr} = D_k,$$

has a solution (X, Y) s.t.

33 $$(X, Y) \in \mathbb{R}[s]^{n_i \times n_i} \times \mathbb{R}[s]^{n_i \times n_o}$$

and

the left fraction (X, Y) is int. pr. with

62 X row-reduced with row degrees r_i, $i \in \underline{n}_i$
and highest row-degree coefficient matrix X_h

if and only if

(a)

63 D_k is row-column-reduced with row and column powers r_i, $i \in \underline{n}_i$, resp. k_j, $j \in \underline{n}_i$,

or equiv. (see criterion (3.3.1.63)),

64 $D_k(s) = \mathrm{diag}[s^{r_i}]_{i=1}^{n_i} D_{k-}(s)\, \mathrm{diag}[s^{k_j}]_{j=1}^{n_i}$,

where $D_{k-} \in \mathbb{R}(s)^{n_i \times n_i}$ is <u>biproper</u> and the highest degree coefficient matrix D_{kh} of D_k satisfies

65 $$D_{kh} = D_{k-}(\infty);$$

(b)

66 for the given D_k, (X,Y) is a polynomial matrix solution of comp. eqn. (32) s.t. the row degrees of Y satisfy

$$\partial_{r_i}[Y] \le r_i \quad \forall i \in \underline{n}_i.$$

Moreover, under conditions (a)-(b) we have that the highest degree coefficient matrices of D_{pr}, X, and D_k are related by

67 $$D_{kh} = X_h\, D_{ph}$$

s.t., if $D_{ph} = I$ and $D_{kh} = I$, then $X_h = I$. ∎

67 <u>Comment</u>. From Comments (44) and statement (62), which is an expansion of (34), the order of the solving compensator C is $\sum_{i=1}^{n_i} r_i$, where the r_i's are the row powers of the chosen matrix D_k (see (63)). We shall see below that <u>if these row powers</u> r_i <u>are sufficiently large</u>, then solutions (X, Y) satisfying (66) will exist. Hence we must choose $\chi(s) = \det D_k(s)$, the characteristic polynomial of Σ, to be of sufficiently large degree for problem COMP to have a solution.

70 <u>Proof of Theorem 61</u>. ⇒ : Since $P \in \mathbb{R}_{p,o}(s)^{n_o \times n_i}$ has an int. pr. r.c.f. (N_{pr}, D_{pr}), with D_{pr} column-reduced, we have

71 $D_{pr} = D_{p-}\ \mathrm{diag}[s^{k_j}]$ and $N_{pr}\ \mathrm{diag}[s^{-k_j}] \in \mathbb{R}_{p,o}(s)^{n_o \times n_i}$,

where $D_{p-} \in \mathbb{R}(s)^{n_i \times n_i}$ is biproper and in (4)

72 $$D_{ph} = D_{p-}(\infty).$$

Since $X^{-1}\,Y \in \mathbb{R}_p(s)^{n_i \times n_o}$ has an int. pr. ℓ.f. (X, Y) with X row-reduced, we have similarly

73 $X = \mathrm{diag}[s^{r_i}]\ X_-$ and $\mathrm{diag}[s^{-r_i}]\ Y \in \mathbb{R}_p(s)^{n_i \times n_o}$,

where $X_- \in \mathbb{R}(s)^{n_i \times n_i}$ is biproper and in (62)

74 $$X_h = X_-(\infty).$$

Hence, by (71)-(74), the comp. eqn. (32) reads

75 $$D_k = \mathrm{diag}[s^{r_i}]\ D_{k-}\ \mathrm{diag}[s^{k_j}],$$

where

76 $D_{k-} := X_-\ D_{p-} + \mathrm{diag}[s^{-r_i}]\ YN_{pr}\ \mathrm{diag}[s^{-k_j}]$ is biproper

with

77 $$D_{kh} := D_{k-}(\infty) = X_-(\infty)\ D_{p-}(\infty) =: X_h\ D_{ph}.$$

Hence (64) follows from (75)-(76), and (66) follows from (73^b). Note also that (77) is (67).

⇐ : We note again that, by the plant assumptions, (71)-(72) hold; moreover, (66) implies (73^b), where the r_j are the row powers of D_k in (64). Using now comp. eqn. (32), we have successively

$$\mathrm{diag}[s^{r_i}]\ D_{k-}\ \mathrm{diag}[s^{k_j}] = XD_{p-}\ \mathrm{diag}[s^{k_j}] + Y\ N_{pr},$$

and

78 $$\mathrm{diag}[s^{-r_i}]\ X = \left\{D_{k-} - (\mathrm{diag}[s^{-r_i}]\ Y)(N_{pr}\ \mathrm{diag}[s^{-k_j}])\right\}\ D_{p-}^{-1}.$$

Note that on RHS of (78) we obtain a product of two biproper factors, the first one being biproper becasue (Diag[s^{-r_i}] Y) is proper and (N_{pr} diag[s^{-k_j}]) is strictly proper by (73^b) and (71^b). Hence

79 $$X = \text{diag}[s^{r_i}]\, X_- \text{ with } X_- \text{ biproper.}$$

Therefore, X is row-reduced with row degrees r_i and highest row-degree coefficient matrix

$$X_h = X_-(\infty).$$

Moreover, since diag[s^{-r_i}] Y is proper by (73^b), it follows that

80 $$X^{-1}Y = X_-^{-1}\,\text{diag}[s^{-r_i}]Y \in \mathbb{R}_p(s)^{n_i \times n_o}.$$

Hence by (79)-(80) we have shown that conditions (a)-(b) imply the existence of a solution s.t. (33) and (62) hold. ∎

Condition (66) of Theorem 61 shows that in order that the solution (X, Y) be int. proper, the row degrees of Y must be suitably bounded. The next theorem gives sufficient conditions for the existence of such solutions.

84 <u>Theorem</u> [Existence of int. proper solutions of comp. eqn. (32)]. Consider the plant specifications (2)-(4) and the polynomial matrix $D_k \in \mathbb{R}[s]^{n_i \times n_i}$.

85 Let ($D_{p\ell}$, $N_{p\ell}$) be any int. pr. ℓ.c.f. of P with $D_{p\ell}$ row-reduced. Then the comp. eqn. (32) has a solution (X, Y) s.t.

33 $$(X, Y) \in \mathbb{R}[s]^{n_i \times n_i} \times \mathbb{R}[s]^{n_i \times n_o}$$

and

34 the ℓ.f. (X, Y) is int. pr. with X row-reduced with row degrees r_i, $i \in \underline{n}_i$,

if

63 $D_k \in \mathbb{R}[s]^{n_i \times n_i}$ is r.c.r. with row powers r_i, $i \in \underline{n}_i$, and column powers k_j, $j \in \underline{n}_i$,

s.t.

86 $r_i \geq \mu - 1 \quad \forall i \in \underline{n}_i$,

where

87 $\mu := \max\{\partial_{ri}[D_{p\ell}],\ i \in \underline{n}_o\}$,

or equiv.

88 μ is the maximal degree of any entry of $D_{p\ell}$ where $D_{p\ell}$ is the left plant denominator in (85).

89 Comments. (a) Note that an int. pr. ℓ.c.f. $(D_{p\ell}, N_{p\ell})$ of P with $D_{p\ell}$ row-reduced is available in (36)-(38). Note also by Comment 2.4.3.33.L, that the row degrees of $D_{p\ell}$ are unique for a given plant P. In fact μ, as given by (88), is the observability index of the plant P [Kai. 1, Sec. 6.4.3, p. 413; Sec. 6.4.6, p. 431]; i.e., let [A, B, C, 0] be any minimal realization of P, then μ is the smallest integer ℓ s.t.

$$\mathrm{rk}\begin{bmatrix} C \\ CA \\ \vdots \\ CA^{\ell-1} \end{bmatrix} = \partial[\det(sI-A)] = \partial[\det D_{p\ell}].$$

or equiv. μ-1 is the minimum number of derivatives of the output needed to reconstruct all states at t = 0.

(b) Theorem 84 tells us that for solving problem COMP a safe choice of D_k is such that (63) and (86) hold. This implies that $\partial[\chi] = \partial[\det D_k]$ $\geq \sum_{i=1}^{n_i} \{k_i + (\mu-1)\} = \partial[\det D_{pr}] + n_i(\mu-1)$. Compare with [Kai.1, Sec. 7.5, p. 535]. This implies also that the order of the compensator satisfies $\partial[\det D_{c\ell}] = \partial[\det X] \geq n_i(\mu-1)$.

(c) Condition (86) may be very conservative (see [Emr.1]): smaller row powers r_i may be acceptable for satisfying the conditions of Theorem 61 characterizing "internally proper" solutions.

91 Proof of Theorem 84. Return to Theorem 39 characterizing all polynomial matrix solutions (X, Y) of eqn. (32) . These solutions, for a D_k satisfying

(63) and (86), are given by

41 $$X = D_k V_{pr} - N_k N_{p\ell},$$

42 $$Y = D_k U_{pr} + N_k D_{p\ell}.$$

Now, in (42), we may consider Y as the result of the division of $D_k U_{pr}$ on the right by $D_{p\ell}$, i.e.,

$$D_k U_{pr} = -N_k D_{p\ell} + Y \quad \text{s.t.} \quad Y D_{p\ell}^{-1} \in \mathbb{R}_{p,o}(s)^{n_i \times n_o}.$$

Hence, by Fact 2.4.3.4.R,

$$\partial_{ci}[Y] < \partial_{ci}[D_{p\ell}] \quad \forall i \in \underline{n}_o.$$

Hence $\forall i \in \underline{n}_i$, $j \in \underline{n}_o$

$$\begin{aligned} \partial_{ri}[Y] &\le \max \partial_{ri}[Y] \\ &= \max \partial_{cj}[Y] < \max \partial_{cj}[D_{p\ell}] \\ &= \max \partial_{rj}[D_{p\ell}] =: \mu. \end{aligned}$$

Therefore, using (86),

$$\partial_{ri}[Y] \le \mu - 1 \le r_i \quad \forall i \in \underline{n}_i.$$

Hence we satisfy condition (66) of Theorem 61. Now, since condition (63) holds by assumption, all conditions of Theorem 61 hold and the comp. eqn. (32) must have a solution (X, Y) s.t. (33)-(34) holds. ∎

93 Summary for design. From a designer's point of view the results above can be summarized as follows:

$$C := D_{c\ell}^{-1} N_{c\ell} \in \mathbb{R}_p(s)^{n_i \times n_o} \text{ solves problem COMP}$$

if and only if

$(D_{c\ell}, N_{c\ell}) := (X, Y)$ is a solution of the comp. eqn.

32 $$XD_{pr} + YN_{pr} = D_k,$$

where

(a)

94 $$\det D_k(s) = \chi(s) \quad \text{with} \quad Z[\chi] \subset \mathring{\mathbb{C}}_-,$$

where i) χ is the prescribed char. poly. of Σ, and ii) D_k has the dynamical interpretation of Fact 30;

(b)

63 D_k is r.c.r. with row powers r_i, $i \in \underline{n}_i$ and column powers k_j, $j \in \underline{n}_i$,

where (i) the row powers must be supplied, (ii) the column powers are given by the plant in (3), and (iii) the highest degree coefficient matrix D_{kh} specifies X_h by (67) and (4);

(c)

66 $$\partial_{ri}[Y] \leq r_i \quad \forall i \in \underline{n}_i.$$

95 <u>Comment</u>. The summary above suggests also the following parametrization of all int. pr. solutions (X, Y) of (32) for a D_k satisfying (94) and (63): $(X, Y) \in \mathbb{R}[s]^{n_i \times n_i} \times \mathbb{R}[s]^{n_i \times n_o}$ must be s.t. (a) Y is given by the conditions

96 $$D_{pr} \text{ divides } D_k - YN_{pr} \text{ on the right},$$

66 $$\partial_{ri}[Y] \leq r_i \quad \forall i \in \underline{n}_i,$$

and (b) X is given by the condition

97 $$D_k - YN_{pr} = XD_{pr}.$$

Since X is determined as a quotient by (97) once Y is given, it follows finally that all internally proper solutions (X, Y) of (32) for the given D_k are parameterized by Y satisfying (96) and (66).

A further transformation of the conditions (a) and (b) is possible. Let

$$n_p := \sum_{j=1}^{n_i} k_j = \partial[\det D_{pr}],$$

$$n_c := \sum_{i=1}^{n_i} r_i,$$

and assume, for reasons of simplicity, that

$$D_{pr} \in \mathbb{R}[s]^{n_i \times n_i} \text{ has } n_p \text{ distinct eigenvalues } p_q,\ q \in \underline{n}_p.$$

Hence D_{pr} has n_p nonzero eigenvectors $\ell_q \in \mathbb{C}^{n_i}$, $q \in \underline{n}_p$ s.t.

$$D_{pr}(p_q)\ell_q = \theta \quad \forall q \in \underline{n}_p.$$

Then, by [Goh.2, Th. 4.2], (96) is equivalent to

$$D_k(p_q)\ell_q = Y(p_q)N_{pr}(p_q)\ell_q \qquad \forall q \in \underline{n}_p,$$

and (66) is equivalent to

$$Y(s) = S(s)Y_c,$$

where

$$S(s) := \text{block diag}\left[[s^{r_i}, s^{r_i-1}, \cdots, 1],\ i \in \underline{n}_i\right] \in \mathbb{R}[s]^{n_i \times (n_c+n_i)},$$

$$Y_c \in \mathbb{R}^{(n_c+n_i)\times n_o}.$$

Note that matrix Y_c has as entries the coefficients of the polynomial entries of $Y(\cdot) \in \mathbb{R}[s]^{n_i \times n_o}$. Finally, let $A \otimes B := [a_{ij}\ B]_{i,j}$ denote the Kronecker product of two matrices and $\sigma[Y_c] \in \mathbb{R}^{n_o(n_c+n_i)}$ denote a vector obtained by stacking the columns of Y_c. Then one obtains finally that conditions (96) and (66) are equivalent to the following <u>linear equation</u> in $\sigma[Y_c]$:

98 $$[D_k(p_q)\ell_q]_{q=1}^{n_p} = [(N_{pr}(p_q)\ell_q)^T \otimes S(p_q)]_{q=1}^{n_p}\ \sigma[Y_c],$$

where the vector on the LHS is of dimension $n_i \cdot n_p$ and the matrix on the RHS has dimension $n_i n_p \times n_o(n_c + n_i)$. Equation (98) parametrizes all Y satisfying (96) and (66) and hence all int. pr. solutions (X, Y) of (32) for the given D_k.

Chapter 7. Asymptotic Tracking

7.1. Introduction

This chapter develops first the theory of asymptotic tracking of a given class of inputs by a unity feedback system, where the forward path consists of a compensator followed by a given plant. Next the theory is used to design a tracking compensator.

Section 2 on theory starts with the analysis (7.2.4) of the tracking error of an exponentially stable unity feedback system. After defining (7.2.16) the class Ψ of inputs to be tracked and describing (7.2.19) the plant assumptions, the problem of asymptotic tracking is defined and discussed (7.2.25 et seg.). This leads (Theorem 7.2.31) to necessary conditions for asymptotic tracking on the compensator-plant forward path: essentially the forward path transfer function must be of full normal rank and may not have any zeros at the input characteristic frequencies. Theorem 7.2.60 develops sufficient conditions for asymptotic tracking: they are (1) exponential stability of the unity feedback system and (2) the factor condition: every entry of the compensator denominator (generated by a right coprime fraction) must have the input characteristic polynomial as a factor. (The latter condition is also known as the "internal model principle.") In Remark 7.2.73 we show that the problem of asymptotic disturbance rejection is almost equivalent to that of asymptotic tracking. In Theorem 7.2.75 we discuss the robustness of asymptotic tracking and disturbance rejection under the conditions of Theorem 7.2.60. Finally, Corollary 7.2.80 reformulates the tracking conditions in a form useful for design.

Section 3 treats the design of a tracking compensator. The problem is described (7.3.1), and the solution is given by an algorithm whose main step consists of solving the tracking compensator equation (7.3.18). This equation has as solution the polynomial matrix numerator and denominator of the tracking compensator. Theorem 7.3.33 describes the class of all polynomial matrix solutions. Finally, Theorem 7.3.45 characterizes internally proper solutions, while Theorem 7.3.55 gives a sufficient condition for their existence.

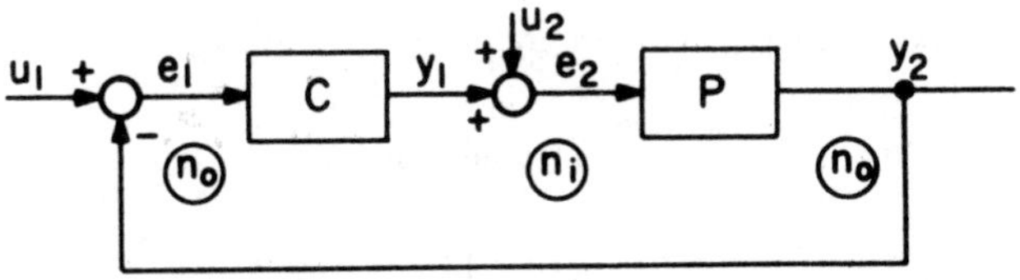

Fig. 4.3.1. The feedback system Σ under consideration.

7.2. Theory of Asymptotic Tracking

We study again feedback system Σ, shown in Fig. 4.3.1. As before P is the plant, C is the compensator, but now, for convenience, we call u_1, e_1, and y_2 resp. the input, error, and output of Σ.

1 Assumption. In feedback system Σ, shown in Fig. 4.3.1, the plant P and compensator C are such that

2 $P \in \mathbb{R}_{p,o}(s)^{n_o \times n_i}$ has an int. pr. ℓ.c.f. $(D_{p\ell}, N_{p\ell})$ and an int. pr. r.c.f. (N_{pr}, D_{pr}),

3 $C \in \mathbb{R}_p(s)^{n_i \times n_o}$ has an int. pr. ℓ.c.f. $(D_{c\ell}, N_{c\ell})$ and an int. pr. r.c.f. (N_{cr}, D_{cr}).

4 Analysis. Under Assumption 1, Theorem 4.3.6 on feedback system stability holds. In particular, with $P = D_{p\ell}^{-1}N_{p\ell}$, $C = N_{cr}D_{cr}^{-1}$, and $u_2 \equiv \theta$, the error e_1 of Σ is described by the well-formed PMD $\mathcal{D}_{e_1u_1} = [D_{p\ell}D_{cr} + N_{p\ell}N_{cr}, D_{p\ell}, D_{cr}, 0]$, i.e., by the equations

5 $$[D_{p\ell}D_{cr} + N_{p\ell}N_{cr}](p)\xi_c(t) = D_{p\ell}(p)u_1(t) \qquad \forall t \geq 0,$$

6 $$e_1(t) = D_{cr}(p)\xi_c(t)$$

where $\xi_c(\cdot)$ is the compensator pseudo-state; moreover, $\xi_p(\cdot) = y_2(\cdot)$, the plant-pseudo-state, is then given by

7 $$\xi_p(t) = -D_{cr}(p)\xi_c(t) + u_1(t) \qquad \forall t \geq 0.$$

Hence, using the char. poly. of Σ, viz.,

8 $$\chi(s) := \det[D_{p\ell}D_{cr} + N_{p\ell}N_{cr}](s)$$

given by (4.3.9), and the input-error transfer function $H_{e_1u_1}$ given by

9 $$H_{e_1u_1} = D_{cr}[D_{p\ell}D_{cr} + N_{p\ell}N_{cr}]^{-1}D_{p\ell} \in \mathbb{R}_p(s)^{n_o \times n_o},$$

it follows that, for every initial value at t = 0- of $\xi_c(\cdot)$ and $\xi_p(\cdot)$ and their derivatives, and for every p. suff. diff. $u_1(\cdot)$ with $u_1^{(j)}(0-) = \theta$ $\forall j = 0, 1, 2, \cdots$, $e_1(\cdot)$ shall be p. suff. diff. on $t \geq 0$ with Laplace transform

10 $$\hat{e}_1(s) = H_{e_1u_1}(s)\hat{u}_1(s) + \chi(s)^{-1}\ m(s),$$

where

11 $$m(s) \in \mathbb{R}[s]^{n_o}$$

is a polynomial vector depending on the initial values $\xi_c^{(j)}(0-)$ for $j = 0, 1, 2, \cdots$, s.t.

12 $$\chi(s)^{-1}m(s) \in \mathbb{R}_{p,o}(s)^{n_o}.$$

Moreover,

13 $$\hat{u}_1 \in \mathbb{R}_{p,o}(s)^{n_o} \Rightarrow \hat{e}_1 \in \mathbb{R}_{p,o}(s)^{n_o},$$ ■

14 Exercise. Show that under Assumption 1, with $u_2 \equiv \theta$, the output y_2 of Σ is described by the PMD $\mathcal{D}_{y_2u_1} := [(D_{c\ell}D_{pr} + N_{c\ell}N_{pr}), N_{c\ell}, N_{pr}, 0]$.

In this section we propose to study the problem of asymptotic tracking under the following additional assumptions

16 Assumption. The class of inputs u_1 of Σ to be tracked is specified in terms of the Laplace transform as follows:

17 $$\Psi := \{\hat{u}_1 = \psi^{-1}\mu : \psi \in \mathbb{R}[s],\ \mu \in \mathbb{R}[s]^{n_o},\ \psi^{-1}\ \mu \in \mathbb{R}_{p,o}(s)^{n_o}\},$$

where ψ is a given monic polynomial and μ is arbitrary; furthermore,

18 $$Z[\psi] \subset \mathbb{C}_+ .$$

19 Assumption. The plant $P \in \mathbb{R}_{p,o}(s)^{n_o \times n_i}$ is such that

20 $$n_o \leq n_i$$

and

21 $$Z[\psi] \cap Z[P] = \phi .$$ ∎

The problem of asymptotic tracking is now described by

25 Definition. We say that feedback system Σ, shown in Fig. 4.3.1, tracks asymptotically every input u_1 of class Ψ iff, with $u_2 \equiv \theta$, for every initial value at $t = 0-$ of the plant and compensator pseudo-states $\xi_p(\cdot)$ and $\xi_c(\cdot)$ and their derivatives, and for every input $u_1 \in \Psi$, with $u_1(t) = \theta_{n_o}$ for $t < 0$, we have that

$$e_1(t) \to \theta_{n_o} \quad \text{as } t \to \infty$$

or equiv.

$$y_2(t) - u_1(t) \to \theta_{n_o} \quad \text{as } t \to \infty .$$ ∎

26 Comment. For brevity, call the inputs $u_1 \in \Psi$ the reference inputs. They are described by (17)-(18) and are characterized as follows: Let

$$\psi(s) := \prod_{\ell=1}^{m} (s - \zeta_\ell)^{k_\ell} ,$$

where ζ_ℓ, for $\ell = 1, \cdots, m$, denotes the distinct roots of $\psi(\cdot)$. Let $1(t)$ denote the unit step function.

U.t.c.

(a)

27 $$u_1 \in \Psi$$

iff

$$u_1(t) = 1(t) \sum_{\ell=1}^{m} \sum_{q=0}^{k_\ell - 1} u_{\ell q} t^q e^{\zeta_\ell t} \qquad \forall t ,$$

where $\forall \ell$, q, $u_{\ell q} \in \mathbb{C}^{n_o}$ is arbitrary except for the fact that $u_{\ell q} = \bar{u}_{pq}$ if $\zeta_\ell = \bar{\zeta}_p \in \mathbb{C}\backslash\mathbb{R}$ and $u_{\ell q} \in \mathbb{R}^{n_o}$ if $\zeta_\ell \in \mathbb{R}$.

(b)

28 $$u_1 \in \Psi$$

iff

$$u_1(t) = \theta_{n_o} \quad \text{for} \quad t < 0$$

and for $t \geq 0$ <u>every component</u> of $u_1(\cdot)$ is a solution of the scalar differential equation

$$\psi(p)x(t) = 0 \quad t \geq 0$$

for arbitrary initial conditions on $x(\cdot)$ and its derivatives at $t = 0$. ∎

In other words, the class Ψ is a set of linear combinations of waveforms of fixed shape.

Well-known waveforms to be tracked are bounded waveforms such as steps and sinusoids, and unbounded waveforms such as ramps, parabolas and increasing exponentials, justifying the requirement $Z[\psi] \subset \mathbb{C}_+$. Note that a bounded waveform going to θ_{n_o} as $t \to \infty$ is always tracked once the feedback system Σ is exp. stable (see Th. 3.3.2.19).

29 <u>Exercise</u>. Consider an exp. stable feedback system Σ satisfying Assumptions 1, 16, and 19. Assume that Σ tracks asymptotically every input of the class Ψ. Let $v_1 : \mathbb{R}_+ \to \mathbb{R}^{n_o}$ be bounded on compact intervals. Let $v_1(t) = \theta_{n_o}$ for $t < 0$, and, for some $u_1 \in \Psi$, $v_1(t) - u_1(t) \to \theta_{n_o}$, as $t \to \infty$. Show that, for any initial conditions at $t = 0-$ on $\xi_c(\cdot)$ and $\xi_p(\cdot)$ and their derivatives, the output due to these initial conditions and to $v_1(\cdot)$ satisfies $y_2(t) - u_1(t) \to \theta_{n_o}$.

30 <u>Comment</u>. The problem of asymptotic tracking occurs in practice in the following contexts.

(a) <u>Regulators</u> (e.g., Manufacturing processes: control of position, velocity, temperature, pH, ...). The problem is to track a set point which may be changed from time to time (hence step inputs).

(b) Servomechanisms. For example, a radar plotting board following an airplane: the horizontal position of an airplane is determined by its polar coordinates with respect to the radar: the angle $\alpha(\cdot)$ (azimuth) and the distance $d(\cdot)$. The requirement is to track $\alpha(\cdot)$ and $d(\cdot)$, two slowly varying functions of time. The idea is to use polynomial interpolation: e.g., $\alpha(t) \simeq \alpha_0 + \alpha_1 t + \alpha_2 t^2$, a parabola, or $\alpha(t) \simeq \alpha_0 + \alpha_1 t + \alpha_2 t^2 + \alpha_3 t^3$, a cubic, and to require that the feedback system tracks any such polynomial sufficiently fast with zero asymptotic error. Note that an arbitrary parabola is a solution of $\dddot{x} = 0$.

A first task is to show that assumptions (20) and (21) are not restrictive.

31 Theorem [Necessary conditions for asymptotic tracking]. Let Assumption 1 hold and let Ψ be the class of reference inputs u_1 described by (17)-(18). Let Σ be exp. stable.

U.t.c., if

Σ tracks asymptotically every input $u_1 \in \Psi$,

then

32 $$\forall\zeta \in Z[\psi] \quad \det[N_{p\ell} N_{cr}](\zeta) \neq 0,$$

whence

33 $$n_o \leq n_i,$$

and

34 $$\forall\zeta \in Z[\psi] \quad \mathrm{rk}[N_{p\ell}](\zeta) = n_o,$$

and

35 $$\forall\zeta \in Z[\psi] \quad \mathrm{rk}[N_{cr}](\zeta) = n_o,$$

or equivalently:

with $PC \in \mathbb{R}_{p,o}(s)^{n_o \times n_o}$ denoting the forward path transfer function,

36 $$Z[PC] \cap Z[\psi] = \phi,$$

whence

37 $$n_o \leq n_i,$$

38 $$Z[P] \cap Z[\psi] = \phi,$$

and

39 $$Z[C] \cap Z[\psi] = \phi.$$

40 Comments. Theorem 31 teaches us that an exp. stable feedback system Σ will asymptotically track every input $u_1 \in \Psi$ only if the forward path transfer function PC has full normal rank n_o and no zeros at the input characteristic frequencies $\zeta \in Z[\psi]$. This implies that $n_o \le n_i$ and that the plant and compensator have no zeros at any $\zeta \in Z[\psi]$. Hence assumptions (20) and (21) are necessary.

42 Proof of Theorem 31. (a) We shall first show the necessity of condition (32). By Assumption 1 feedback system Σ satisfies the assumptions of Theorem 4.3.6 and is exp. stable. Hence, with Definition 4.2.66, Σ is a well-formed PMD $\mathcal{D}_g$ which is exp. stable with char. poly.

43 $$\chi(s) := \det[D_{p\ell}D_{cr} + N_{p\ell}N_{cr}](s)$$

s.t.

44 $$Z[\chi] \subset \mathring{\mathbb{C}}_-.$$

By assumption also Σ tracks asymptotically every input $u_1 \in \Psi$ and, in particular, every input of the form

45 $$u_1(t) := u_o e^{\zeta t} \quad \text{where} \quad \zeta \in Z[\psi] \quad \text{and} \quad u_o \in \mathbb{C}^{n_o}.$$

Hence, with $u_2 \equiv \theta$, and by the exp. stability of Σ: $\forall \zeta \in Z[\psi]$, $\forall u_o \in \mathbb{C}^{n_o}$, $\forall t \ge 0$

46 $$y_2(t) = u_o e^{\zeta t} + \sum_i \sum_j y_{ij} t^j e^{\lambda_i t}$$

47 $$e_2(t) = y_1(t) = H_{y_1u_1}(\zeta) u_o e^{\zeta t} + \sum_i \sum_j e_{ij} t^j e^{\lambda_i t},$$

where (a) the λ_i's are (closed-loop) eigenvalues of Σ, whence $\mathrm{Re}\,\lambda_i < 0\ \forall i$ and (b) the y_{ij}'s and e_{ij}'s are constant vectors which depend on the initial conditions and the input.

Note further that

48 (i) $$\forall \zeta \in Z[\psi] \quad H_{y_1u_1}(\zeta) = [N_{cr}(D_{p\ell}D_{cr} + N_{p\ell}N_{cr})^{-1}D_{p\ell}](\zeta) \in \mathbb{C}^{n_i \times n_o}$$

because of (43)-(44) and (18), and

49 (ii) $$D_{p\ell}(p)y_2(t) = N_{p\ell}(p)e_2(t) \quad \forall t \geq 0,$$

where the latter equation is due to the plant PMD $\mathcal{D}_p = [D_{p\ell}, N_{p\ell}, I, 0]$ with $y_2(\cdot) = \xi_p(\cdot)$. Therefore, using (46)-(49) and letting $t \to \infty$, we obtain

$$\forall \zeta \in Z[\psi] \quad \forall u_o \in \mathbb{C}^{n_o}$$

50 $$D_{p\ell}(\zeta)u_o = (N_{p\ell}N_{cr})(\zeta)\ \{(D_{p\ell}D_{cr} + N_{p\ell}N_{cr})^{-1}(\zeta)\}\ D_{p\ell}(\zeta)u_o,$$

an equation which is well defined because of (43)-(44) and (18). Assume now for the purpose of a contradiction that (32) does not hold, i.e.,

51 $$\exists\ \bar{\zeta} \in Z[\psi] \quad \text{s.t.} \quad \det[N_{p\ell}N_{cr}](\bar{\zeta}) = 0.$$

or equiv.

52 $$\exists\ \bar{\zeta} \in Z[\psi] \text{ and } \exists \text{ a nonzero } \gamma \in \mathbb{C}^{n_o} \text{ s.t. } \gamma^*(N_{p\ell}N_{cr})(\bar{\zeta}) = \theta^*.$$

Hence in view of (50) and (52),

53 $$\forall u_o \in \mathbb{C}^{n_o} \quad \gamma^*D_{p\ell}(\bar{\zeta})\ u_o = 0.$$

Note now that, since Σ is exp. stable, by the analysis done in (4), the PMD $\mathcal{D}_{e_1u_1} = [D_{p\ell}D_{cr} + N_{p\ell}N_{cr}, D_{p\ell}, D_{cr}, 0]$ is exp. stable, and therefore does not have any unstable hidden modes. Hence, using some e.o.'s, we obtain

54 $$\text{rk}[N_{p\ell}N_{cr} \,\vdots\, D_{p\ell}](s) = n_o \quad \forall s \in \mathbb{C}_+,$$

55 $$\text{rk}\begin{bmatrix} N_{p\ell}\ N_{cr} \\ \hline D_{cr} \end{bmatrix}(s) = n_o \quad \forall s \in \mathbb{C}_+,$$

It follows now by (52), (54), and $\bar{\zeta} \in Z[\psi] \subset \mathbb{C}_+$ that

$$\eta^* := \gamma^* D_{p\ell}(\bar{\zeta}) \neq \theta^*.$$

However, setting $u_o = \eta$ in (53), we must also have

$$\eta^*\eta = 0, \text{ equiv. } \eta = \theta.$$

Hence we have reached a contradiction and assumption (51) is false, or equiv. (32) holds.

(b) We now establish the necessity of (33)-(35). We have that $N_{p\ell} \in \mathbb{R}[s]^{n_o \times n_i}$ and $N_{cr} \in \mathbb{R}[s]^{n_i \times n_o}$, whence by (32) and Sylvester's rule $\forall \zeta \in Z[\psi]$

$$n_o = rk[N_{p\ell}N_{cr}](\zeta) \leq \min\{rk[N_{p\ell}](\zeta), rk[N_{cr}](\zeta)\}$$

$$\leq \min(n_o, n_i) \leq n_i.$$

Hence (33)-(35) hold.

(c) The equivalence between (32) and (36) follows from the fact that by (54)-(55) $D_{p\ell}^{-1} N_{p\ell} N_{cr} D_{cr}^{-1}$ is a left-right fraction of PC with no unstable hidden modes, hence $Z[PC] \cap \mathbb{C}_+ = Z[N_{p\ell}N_{cr}] \cap \mathbb{C}_+$.

(d) Conditions (38) and (39) follow from (33)-(35) and Theorem 2.4.4.6. ■

56 <u>Exercise</u>. Consider feedback system Σ (see Fig. 4.3.1), which satisfies Assumption 1. Let (N_r, D_r) be an int. pr. r.c.f. of PC; hence $H_{e_1u_1} = D_r(D_r + N_r)^{-1}$. Assume that Σ is exp. stable. Show that if Σ tracks asymptotically every input of class Ψ defined in (16), then

(a) $H_{e_1u_1}(s) = \psi(s)R(s)$ for some $R(s) \in \mathbb{R}_p(s)^{n_o \times n_o}$ s.t. $R(s)$ is analytic in $\mathbb{C}_+$;

(b) $D_r(s) = \psi(s)D(s)$ for some $D(s) \in \mathbb{R}[s]^{n_o \times n_o}$;

(c) $\det N_r(\zeta) \neq 0$, $\forall \zeta \in Z[\psi]$.

(Hint: Note Exercise 3.3.2.28.) ■

We have now the main result.

60 <u>Theorem</u> [Sufficient condition for asymptotic tracking]. Consider feedback system Σ, shown in Fig. 4.3.1, under the system Assumptions 1, the input Assumptions 16, and the plant Assumptions 19.

U.t.c., <u>if</u>

(a)

61 feedback system Σ is exp. stable

or equiv.

62 $\chi(s) := \det[D_{p\ell}D_{cr} + N_{p\ell}N_{cr}](s)$ is s.t. $Z[\chi] \subset \mathring{\mathbb{C}}_-$;

(b)

63 $\psi \in \mathbb{R}[s]$ is a factor of every entry of D_{cr},

or equiv.

64 $D_{cr}(s) = \psi(s)D_c(s)$ for some $D_c \in \mathbb{R}[s]^{n_o \times n_o}$;

<u>then</u>

65 feedback system Σ tracks asymptotically every input $u_1 \in \Psi$.

66 <u>Comment</u>. Condition (63)⇔(64) is known as the "internal model principle" and states that the compensator $C = N_{cr}D_{cr}^{-1} = N_{cr}D_c^{-1}(\psi I_{n_o})^{-1}$ may be viewed as a series connection of two blocks, the first block "reproducing the input dynamics" in n_o parallel uncoupled channels. In particular, if we want to track steps, then one integrator must be present in every channel.

68 <u>Proof of Theorem 60</u>. Under Assumption 1, Theorem 4.3.6 on feedback system stability holds. Hence (62) is a necessary and sufficient condition for exp. stability. By the analysis done in (4), with $u_2 \equiv \theta$, for every initial value at t = 0- of the plant and compensator pseudo-states $\xi_p(\cdot)$ resp. $\xi_c(\cdot)$ and their derivatives and for every p. suff. diff. $u_1(\cdot)$ with $u_1^{(j)}(0-) = \theta$ $\forall j = 0, 1, 2, \cdots$, the error $e_1(\cdot)$ shall be p. suff. diff. on $t \geq 0$ with Laplace transform

10 $$\hat{e}_1(s) = H_{e_1u_1}(s)\hat{u}_1(s) + \chi(s)^{-1}m(s),$$

where

11 $$m(s) \in \mathbb{R}[s]^{n_o}$$

is a polynomial vector depending on the initial values $\xi_c^{(j)}(0-)$ for

$j = 0, 1, 2, \cdots$ s.t.

12 $$\chi(s)^{-1}\, m(s) \in \mathbb{R}_{p,o}(s)^{n_o}.$$

Moreover,

9 $$H_{e_1u_1} = D_{cr}[D_{p\ell}D_{cr} + N_{p\ell}N_{cr}]^{-1}D_{p\ell} \in \mathbb{R}_p(s)^{n_o \times n_o}$$

and

13 $$\hat{u}_1 \in \mathbb{R}_{p,o}(s)^{n_o} \Rightarrow \hat{e}_1 \in \mathbb{R}_{p,o}(s)^{n_o}.$$

Now using (63)⇔(64), we obtain

69 $$H_{e_1u_1} = D_c[D_{p\ell}D_{cr} + N_{p\ell}N_{cr}]^{-1}D_{p\ell}\psi \in \mathbb{R}_p(s)^{n_o \times n_o}$$

since the <u>scalar</u> polynomial ψ commutes with any matrix. Pick any $u_1 \in \Psi$, whence by (17)

70 $$\hat{u}_1(s) = \psi(s)^{-1}\mu(s) \in \mathbb{R}_{p,o}(s)^{n_o}.$$

Hence the substitution of (69)-(70) into (10), and (12)-(13), (61)⇔(62) result in

$$\hat{e}_1(s) = D_c(s)[D_{p\ell}D_{cr} + N_{p\ell}N_{cr}](s)^{-1}D_{p\ell}(s)\mu(s) + \chi(s)^{-1}m(s)$$

s.t.

$$\hat{e}_1 \in \mathbb{R}_{p,o}(s)^{n_o} \text{ and } P[\hat{e}_1] \subset \mathring{\mathbb{C}}_-.$$

As a consequence $e_1(t) \to 0$ as $t \to \infty$, and feedback Σ tracks asymptotically every input u_1 of class Ψ. ■

71 <u>Exercise</u>. Let Assumptions 1, 16, and 19 hold. Let $n_i = n_o =: n$. Show that if $Z[\det(D_{c\ell}D_{pr} + N_{c\ell}N_{pr})(s)] \subset \mathring{\mathbb{C}}_-$ and if $D_{c\ell}D_{pr} = \psi\, D_r$ for some $D_r \in \mathbb{R}[s]^{n\times n}$, then Σ tracks asymptotically every input $\in \Psi$.
(Hint: (i) $H_{y_2u_1} = [I + \psi N_{c\ell}^{-1}D_rN_{pr}^{-1}]$ is valid in some small neighborhood N of $Z[\psi]$. (ii) Write $H_{y_2u_1}$ from (i) as $H_{y_2u_1} = [I + \psi R(s)]^{-1}$ and for N sufficiently small $H_{y_2u_1}(s) = I - \psi(s)R(s) + \psi(s)^2R(s)^2 - \cdots$. (iii)$\cdots$.)

73 Remark: additive plant-output disturbance rejection. Assume that in feedback system Σ, shown in Fig. 4.3.1, plant-output disturbances d_o occur as shown in Fig. 1. Assume also that these disturbances are of class Ψ described by (17)-(18). Note that now

74 $$H_{y_2 d_o} = H_{e_1 u_1} = (I + PC)^{-1}.$$

It follows then, by an analysis similar to the proof of Theorem 60, that, under the assumptions and conditions of Theorem 60, with $u_1 \equiv \theta$ and $u_2 \equiv \theta$. for every initial value at t = 0- of the plant and compensator pseudo-states $\xi_p(\cdot)$ resp. $\xi_c(\cdot)$ and their derivatives, and for every disturbance $d_o \in \Psi$ with $d_o(t) = \theta$ for $t < 0$, $y_2(t) \to \theta$ as $t \to \infty$. Hence under the assumptions and conditions of Theorem 60, feedback Σ will asymptotically track every input $u_1 \in \Psi$ and asymptotically reject every plant-output disturbance $d_o \in \Psi$.

75 Theorem [Robustness of asymptotic tracking and disturbance rejection]. Assume that the feedback system Σ satisfies all the assumptions stated in Theorem 60. Consider arbitrary changes in the plant P, more precisely $N_{p\ell} \to \tilde{N}_{p\ell}$, $D_{p\ell} \to \tilde{D}_{p\ell}$ s.t. $(\tilde{D}_{p\ell}, \tilde{N}_{p\ell})$ is an int. pr. ℓ.c.f., and arbitrary changes in the compensator C, more precisely, $N_{cr} \to \tilde{N}_{cr}$, $D_c \to \tilde{D}_c$ s.t. $(\tilde{N}_{cr}, \psi\tilde{D}_c)$ is an int. pr. r.c.f. Assume further that the exp. stability of Σ is preserved. U.t.c. the perturbed system $\tilde{\Sigma}$ has the asymptotic tracking properties (stated in Theorem 60) and the disturbance rejection properties (stated in Remark 73). ∎

Proof. Follows the same steps as the proof of Theorem 60. ∎

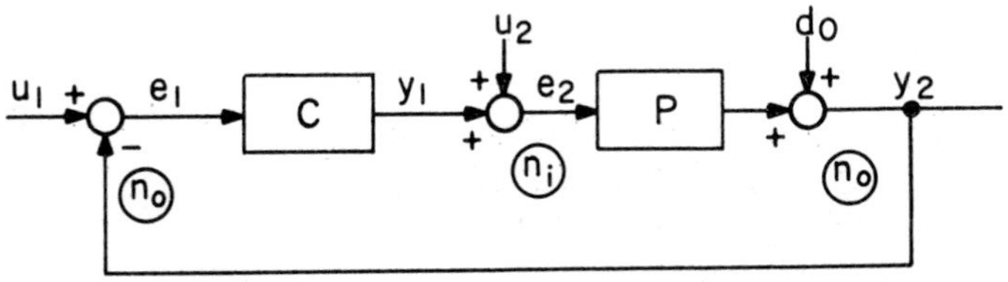

Fig. 1. Feedback system Σ with plant-output disturbances.

76 Comments. (a) The presence of the factor ψ in every element of D_{cr} (see (64)) is the crucial element in robust asymptotic tracking and disturbance rejection: note that P, D_c, and N_{cr} may undergo large perturbations (which must of course preserve exp. stability of Σ).
(b) The effectiveness of this factor ψ is made obvious if we refer to PMD $\mathcal{D}_{e_1u_1}$ and eqn. (5) and (6). For good intuitive reasons, the zeros of ψ have been called blocking zeros of $H_{e_1u_1}$ (see the scalar factor ψ in eqn. (69)).

80 Corollary. [Tracking for compensator design]. Consider feedback system Σ, shown in Fig. 4.3.1, where

81 (1) $n_o = n_i =: n$,

82 (2) $P \in \mathbb{R}_{p,o}(s)^{n\times n}$ has an int. pr. r.c.f. (N_{pr}, D_{pr}),

83 (3) $C \in \mathbb{R}_p(s)^{n\times n}$ has an int. pr. ℓ.f. $(D_{c\ell}, N_{c\ell})$,

84 (4) the input-assumption 16 and plant-assumptions 19 hold.
U.t.c., if

85 (a) feedback system Σ is exp. stable,

or equiv.

86 $$\chi(s) := \det[D_{c\ell}D_{pr} + N_{c\ell}N_{pr}](s) \text{ is s.t. } Z[\chi] \subset \mathring{\mathbb{C}}_-,$$

87 (b) ψ is a factor of every entry of $D_{c\ell}$,

or equiv.

88 $D_{c\ell}(s) = \psi(s)D_c(s)$ for some $D_c \in \mathbb{R}[s]^{n\times n}$,

then

89 feedback system Σ tracks asymptotically every input $u_1 \in \Psi$.

90 Comments (a) By choosing $P = N_{pr}\, D_{pr}^{-1}$ and $C = D_{c\ell}^{-1}\, N_{c\ell}$, one considers implicitly the I/O map

91 $$H_{y_2u_1} = N_{pr}[D_{c\ell}D_{pr} + N_{c\ell}N_{pr}]^{-1}N_{c\ell}.$$

(b) Note also that the corollary does not require C to have a coprime fraction: this is useful because the proposed compensator design does not always deliver C as a coprime fraction. It will also follow from the analysis below that the factor condition (88) will be preserved after canceling common left factors.

(c) Since $Z[\psi] \subset \mathbb{C}_+$ it follows from the stability requirement (86) and the factor condition (88) that

92 $$\det N_{c\ell}(\zeta) \neq 0 \quad \text{and} \quad \det N_{pr}(\zeta) \neq 0 \quad \forall \zeta \in Z[\psi].$$

We start the proof of Corollary 80 with three lemmas.

95 Lemma. Let D and $\bar{D}$ be two equivalent nonsingular polynomial matrices belonging to $\mathbb{R}[s]^{n\times n}$. Let $\psi \in \mathbb{R}[s]$ and let $\psi|D$ denote the fact that ψ is a factor of every entry of D.

U.t.c.

$$\psi|D \Leftrightarrow \psi|\bar{D}.$$

Proof. Since D and $\bar{D}$ are equivalent, $\exists$ unimodular matrices L and R in $\mathbb{R}[s]^{n\times n}$ s.t. $D = L\,\bar{D}\,R$ and $L^{-1}\,D\,R^{-1} = \bar{D}$, where L, L^{-1}, R, R^{-1} are polynomial matrices. Hence, if $D = \psi\tilde{D}$ for some $\tilde{D} \in \mathbb{R}[s]^{n\times n}$, then $\bar{D} = L^{-1}\psi\tilde{D}\,R^{-1}$, whence $\psi|D \Rightarrow \psi|\bar{D}$. Similarly, if $\bar{D} = \psi\check{D}$ for some $\check{D} \in \mathbb{R}[s]^{n\times n}$, then $D = L\psi\check{D}R$, whence $\psi|\bar{D} \Rightarrow \psi|D$. ∎

96 Lemma. Let $(D_{c\ell}, N_{c\ell})$ be any ℓ.f. of $C \in \mathbb{R}(s)^{n\times n}$. Let $\psi \in \mathbb{R}[s]$ s.t.

97 $$\det N_{c\ell}(\zeta) \neq 0 \quad \forall \zeta \in Z[\psi].$$

Let (N_{cr}, D_{cr}) be any r.c.f. of $C \in \mathbb{R}(s)^{n\times n}$.

U.t.c.

98 $$\psi|D_{c\ell} \Rightarrow \psi|D_{cr}.$$

Proof. Let $(\bar{D}_{c\ell}, \bar{N}_{c\ell})$ be any ℓ.c.f. of $C \in \mathbb{R}(s)^{n\times n}$. Now since (N_{cr}, D_{cr}) is a r.c.f. of C and by Theorem 2.4.2.41 D_{cr} and $\bar{D}_{c\ell}$ are equivalent, it follows by Lemma 85 that $\psi|D_{cr} \Leftrightarrow \psi|\bar{D}_{c\ell}$. Hence (98) holds if

99 $$\psi|D_{c\ell} \Rightarrow \psi|\bar{D}_{c\ell}.$$

We show now that (99) holds.

Now if $\psi | D_{c\ell}$, then $D_{c\ell} = \psi D_c$ for some $D_c \in \mathbb{R}[s]^{n\times n}$. Now let $\tilde{L} \in \mathbb{R}[s]^{n\times n}$ be a g.c.ℓ.d. of $(D_c, N_{c\ell})$, whence $\exists\ \tilde{D}_c$ and $\tilde{N}_{c\ell}$ in $\mathbb{R}[s]^{n\times n}$ s.t.

100 $$D_c = \tilde{L}\,\tilde{D}_c,\ N_{c\ell} = \tilde{L}\,\tilde{N}_{c\ell},$$

101 and $$(\tilde{D}_c, \tilde{N}_{c\ell}) \text{ is } \ell.\text{c.}$$

Note now that by (100) and (97) $\det \tilde{N}_{c\ell}(\zeta) \neq 0 \quad \forall \zeta \in Z[\psi]$, whence by (101), $(\psi\tilde{D}_c, \tilde{N}_{c\ell})$ is ℓ.c. Moreover, since, by (100), $D_{c\ell} = \psi D_c = \tilde{L}\psi\tilde{D}_c$ and $N_{c\ell} = \tilde{L}\tilde{N}_{c\ell}$, we have $C = D_{c\ell}^{-1} N_{c\ell} = (\psi\tilde{D}_c)^{-1}\tilde{N}_{c\ell}$. Hence C has a ℓ.c.f. $(\psi\tilde{D}_c, \tilde{N}_{c\ell})$. Now, since $(\bar{D}_{c\ell}, \bar{N}_{c\ell})$ is also a ℓ.c.f. of C, it follows by Theorem 2.4.1.17.L that there exists a unimodular matrix L s.t. $L\psi\tilde{D}_c = \bar{D}_{c\ell}$. Therefore, $\psi | \bar{D}_{c\ell}$ and (99) holds. ¤

103 <u>Lemma</u>. Consider feedback system Σ, shown in Fig. 4.3.1, where
(a) $P \in \mathbb{R}_{p,o}(s)^{n_o\times n_i}$ has an int. pr. ℓ.c.f. $(D_{p\ell}, N_{p\ell})$ and an int. pr. r.c.f. (N_{pr}, D_{pr}),
(c) $C \in \mathbb{R}_p(s)^{n_i\times n_o}$ has an int. pr. ℓ.f. $(D_{c\ell}, N_{c\ell})$ and an int. pr. r.c.f. (N_{cr}, D_{cr}).
U.t.c.

104 $$\det[D_{p\ell}\,D_{cr} + N_{p\ell}\,N_{cr}] \text{ is a factor of } \det[D_{c\ell}D_{pr} + N_{c\ell}N_{pr}].$$

<u>Proof</u>. Let $L \in E(\mathbb{R}[s])$ be any g.c.ℓ.d. of $(D_{c\ell}, N_{c\ell})$, whence there exist $\bar{D}_{c\ell}$ and $\bar{N}_{c\ell} \in E(\mathbb{R}[s])$ s.t. $D_{c\ell} = L\,\bar{D}_{c\ell}$, $N_{c\ell} = L\,\bar{N}_{c\ell}$ with $(\bar{D}_{c\ell}, \bar{N}_{c\ell})$ an int. pr. ℓ.c.f. of C. Hence,

105 $$\det[D_{c\ell}D_{pr} + N_{c\ell}N_{pr}] = \det L\,\det[\bar{D}_{c\ell}D_{pr} + \bar{N}_{c\ell}N_{pr}].$$

Observe that, with $P = N_{pr}\,D_{pr}^{-1} = D_{p\ell}^{-1}\,N_{p\ell}$ and $C = N_{cr}\,D_{cr}^{-1} = \bar{D}_{c\ell}^{-1}\,\bar{N}_{c\ell}$, the assumptions of Theorem 4.3.6 are satisfied, whence

106 $$\det[\bar{D}_{c\ell}D_{pr} + \bar{N}_{c\ell}N_{pr}] \sim \det[D_{p\ell}D_{cr} + N_{p\ell}N_{cr}].$$

Hence, in view of (105) and (106), (104) follows. ∎

110 Proof of Corollary 80. Notice that by assumptions (81)-(83) we satisfy the assumptions of feedback-stability Corollary 4.3.26. Hence (86) is a necessary and sufficient condition for the exp. stability of Σ. Therefore, under (81)-(83) and (86), the feedback system Σ is exp. stable. Moreover, with $u_2 \equiv \theta$, the error $e_1(\cdot)$ of Σ is described by the well-formed and exp. stable PMD $\mathcal{D}_{e_1u_1} = [D_{c\ell}D_{pr} + N_{c\ell}N_{pr}, N_{c\ell}, -N_{pr}, I]$ by the equations

$$\begin{aligned} [D_{c\ell}D_{pr} + N_{c\ell}N_{pr}](p)\xi_p(t) &= N_{c\ell}(p)u_1(t) \\ e_1(t) &= -N_{pr}(p)\xi_p(t) + u_1(t) \end{aligned} \qquad \forall t \geq 0,$$

where $\xi_p(\cdot)$ is the plant pseudo-state. Moreover, $y_1(\cdot) = \xi_c(\cdot)$, viz., the compensator-pseudo-state is given by

$$\xi_c(t) = D_{pr}(p)\xi_p(t) \qquad \forall t \geq 0.$$

Finally, the input-error transfer function $H_{e_1u_1}$ is given by

$$H_{e_1u_1} = I - H_{y_2u_1} = I - N_{pr}\,[D_{c\ell}D_{pr} + N_{c\ell}N_{pr}]^{-1}\,N_{c\ell} \in \mathbb{R}_p(s)^{n\times n}.$$

Let now $\hat{u}_1 = \psi^{-1}\mu \in \Psi$, where Ψ is given by (17)-(18); then, for every initial value at $t = 0-$ of $\xi_p(\cdot)$ and $\xi_c(\cdot)$ and their derivatives, the Laplace transform of the error $e_1(\cdot)$ of Σ is given by

111 $$\hat{e}_1 = H_{e_1u_1}\psi^{-1}\mu + \chi^{-1}m \in \mathbb{R}_{p,o}(s)^n,$$

where χ is the char. poly. (86) and $m \in \mathbb{R}[s]^n$ is a polynomial vector depending on the initial values at $t = 0-$ of $\xi_p(\cdot)$ and its derivatives.

Now let $(D_{p\ell}, N_{p\ell})$ be an int. pr. ℓ.c.f. of P and (N_{cr}, D_{cr}) an intr. r.c.f of C, then

112 $$H_{e_1u_1} = D_{cr}\,[D_{p\ell}D_{cr} + N_{p\ell}N_{cr}]^{-1}\,D_{p\ell} \in \mathbb{R}_p(s)^{n\times n},$$

where we note that because of Lemma 103,

113 $\det[D_{p\ell}D_{cr} + N_{p\ell}N_{cr}]$ is a factor of $\chi(s)$ in (86).

Moreover, since by stability condition (86) and factor condition (88)

$\det N_{c\ell}(\zeta) \neq 0 \quad \forall \zeta \in Z[\psi] \subset \mathbb{C}_+$, we have by Lemma 96 that

114 $$D_{cr} = \psi \tilde{D}_c \text{ for some } \tilde{D}_c \in \mathbb{R}[s]^{n\times n}.$$

Hence in view of (86), (111)-(114) we have

$$\hat{e}_1 = \tilde{D}_c[D_{p\ell}D_{cr} + N_{p\ell}N_{cr}]^{-1}D_{p\ell}\,\mu + \chi^{-1}m \in \mathbb{R}_{p,o}(s)^n,$$

where $P[\hat{e}_1] \subset \mathring{\mathbb{C}}_-$. Therefore, $\lim_{t\to\infty} e_1(t) = \theta$ and feedback system Σ tracks asymptotically every input u_1 of class Ψ. ∎

7.3. The Tracking Compensator Problem

1 Problem TRC. We are given

(a) a class of reference inputs u_1 to be tracked, specified in terms of the Laplace transform by

2 $$\Psi := \{\hat{u}_1 = \psi^{-1}\mu : \psi \in \mathbb{R}[s],\ \mu \in \mathbb{R}[s]^n,\ \psi^{-1}\mu \in \mathbb{R}_{p,o}(s)^n\},$$

where ψ is a given monic polynomial and μ is arbitrary; furthermore,

3 $$Z[\psi] \subset \mathbb{C}_+.$$

(b) a square (i.e., $n_o = n_i =: n$) plant P s.t.

4 $P \in \mathbb{R}_{p,o}(s)^{n\times n}$ has an int. pr. r.c.f. (N_{pr}, D_{pr}) s.t.

5 D_{pr} is column reduced with column degrees k_j, $j \in \underline{n}$,
and

6 with highest column degree coefficient matrix $D_{ph} = I_n$;

7 $Z[P] \cap Z[\psi] = \phi$ or equiv. $\det[N_{pr}](\zeta) \neq 0 \quad \forall \zeta \in Z[\psi]$.

We must find a compensator C s.t.

8 $C \in \mathbb{R}_p(s)^{n\times n}$ has an int. pr. ℓ.f. $(D_{c\ell}, N_{c\ell})$

with

9 $D_{c\ell}$ row reduced,

s.t.

10 the feedback system Σ, shown in Fig. 4.3.1, is exp. stable with prescribed closed-loop characteristic polynomial $\chi(s)$ and

11 the feedback system Σ tracks asymptotically every input u_1 of class Ψ. ■

In view of Corollary 7.2.80, solutions of problem TRC are obtained by the following

15 Algorithm [Solution of problem TRC]

Step 1. Pick $\chi(s) \in \mathbb{R}[s]$ s.t.

16 $$Z[\chi] \subset \mathring{\mathbb{C}}_-.$$

Step 2. Pick $D_k \in \mathbb{R}[s]^{n\times n}$ s.t.

17 (a) $\det D_k(s) = \chi(s)$;

(b) the tracking compensator equation

18 $$X\psi D_{pr} + YN_{pr} = D_k$$

has a solution (X, Y) s.t.

19 $(X, Y) \in \mathbb{R}[s]^{n\times n} \times \mathbb{R}[s]^{n\times n}$

and

20 $(\psi X, Y)$ is an int. pr. ℓ.f. with ψX row reduced.

Step 3. Pick a solution (X, Y) of the tracking comp. eqn. (18) s.t. (19)-(20) holds.

Set

21 $$D_{c\ell} := \psi X \quad N_{c\ell} := Y$$

22 $$C := D_{c\ell}^{-1} N_{c\ell}.$$ ■

23 Comments. (a) C given by (22) is a solution of problem TRC. Indeed, since $\chi(s) = \det[D_{c\ell}D_{pr} + N_{c\ell}N_{pr}](s)$, $Z[\chi] \subset \mathring{\mathbb{C}}_-$, and $D_{c\ell} = \psi X =: \psi D_c$ for some $D_c \in \mathbb{R}[s]^{n\times n}$, all conditions of Corollary 7.2.80 are met.
(b) Matrix D_k has the dynamical interpretation of Fact 6.2.30; the tracking comp. eqn. (18) is similar to comp. eqn. (6.2.32) since it will turn out that $(N_{pr}, \psi D_{pr})$ is an int. pr. r.c.f. of $\psi^{-1}P \in \mathbb{R}_{p,o}(s)^{n\times n}$.

Solving eqn. (18) for the tracking compensator. In design the input class Ψ (2)-(3) and the plant data (4)-(7) are given.

26 Fact. Given the data (2)-(7), we have that $(N_{pr}, \psi D_{pr})$ is an int. pr. r.c.f. of $\psi^{-1}P \in \mathbb{R}_{p,o}(s)^{n\times n}$.

Proof. By (4)-(5), $P \in \mathbb{R}_{p,o}(s)^{n\times n}$ has an int. pr. r.c.f. (N_{pr}, D_{pr}), with D_{pr} column reduced, and by (7) $\det N_{pr}(\zeta) \neq 0$ $\forall \zeta \in Z[\psi]$. Therefore,

$$\mathrm{rk}\begin{bmatrix} \psi D_{pr} \\ \hline N_{pr} \end{bmatrix}(s) = \mathrm{rk}\left[\begin{array}{c|c} \psi I & 0 \\ \hline 0 & I \end{array}\right]\begin{bmatrix} D_{pr} \\ \hline N_{pr} \end{bmatrix}(s) = n \quad \forall s \in \mathbb{C}\backslash Z[\psi]$$

and $\forall \zeta \in Z[\psi]$. Hence $(N_{pr}, \psi D_{pr})$ is a r.c.f. of $\psi^{-1}P \in \mathbb{R}_{p,o}(s)^{n\times n}$. Moreover, this fraction is int. proper since ψD_{pr} is column reduced. ∎

Note now that since, by Fact 26, $(N_{pr}, \psi D_{pr})$ is a r.c.f. of $\psi^{-1}P$, Theorem 2.4.1.25.R delivers a generalized Bezout identity: there exist six polynomial matrices

27 $$U_r,\ V_r,\ \psi D_{p\ell},\ N_{p\ell},\ U_\ell,\ V_\ell \in E(\mathbb{R}[s])$$

s.t.

28 $$\left[\begin{array}{c|c} V_r & U_r \\ \hline -N_{p\ell} & \psi D_{p\ell} \end{array}\right]\left[\begin{array}{c|c} \psi D_{pr} & -U_\ell \\ \hline N_{pr} & V_\ell \end{array}\right] = \left[\begin{array}{c|c} I_n & 0 \\ \hline 0 & I_n \end{array}\right].$$

Moreover,

29 $$(\psi D_{p\ell}, N_{p\ell}) \text{ is a } \ell.\text{c.f. of } \psi^{-1}P,$$

30 $$(D_{p\ell}, N_{p\ell}) \text{ is a } \ell.\text{c.f. of } P,$$

and without loss of generality

31 $\psi D_{p\ell}$ and $D_{p\ell}$ are row reduced.

Proof. If we first replace $\psi D_{p\ell}$ by $D_\ell \in \mathbb{R}[s]^{n\times n}$, then the existence of the resulting six matrices, listed in (27), s.t. (28)-(29) hold, is due to Theorem 2.4.1.25.R. Note in particular that one obtains $(N_{pr}, \psi D_{pr})$ is a r.c.f. and $(D_\ell, N_{p\ell})$ is a ℓ.c.f. of $\psi^{-1}P$, where the denominators ψD_{pr} and D_ℓ are equivalent polynomial matrices (observe that $\psi^{-1}P \in \mathbb{R}(s)^{n\times n}$ and use Theorem 2.4.2.41). Hence, by Lemma 7.2.95, $D_\ell = \psi D_{p\ell}$ for some $D_{p\ell} \in \mathbb{R}[s]^{n\times n}$. This observation justifies the existence of the six matrices (27) s.t. (28)-(31) hold. In particular, (30) holds because $\det N_{p\ell}(\zeta) \neq 0\ \forall\zeta \in Z[\psi]$ (note that the first matrix on the LHS of (28) is unimodular). ∎

33 Theorem [Polynomial solutions of the tracking comp. eqn. (18)]. Consider eqn. (18), where (a) $P = N_{pr} D_{pr}^{-1}$ satisfies (4)-(7), (b) $\psi \in \mathbb{R}[s]$ is a monic polynomial s.t. (3) holds, and (c) $D_k \in \mathbb{R}[s]^{n\times n}$ is s.t. (16)-(17) holds. U.t.c.

34 $(X, Y) \in \mathbb{R}[s]^{n\times n} \times \mathbb{R}[s]^{n\times n}$ is a solution of (18)

$\Leftrightarrow$

$\exists\ N_k \in \mathbb{R}[s]^{n\times n}$ s.t.

35 $X = D_k V_r - N_k N_{p\ell}$

36 $Y = D_k U_r + N_k \psi D_{p\ell}$,

where U_r, V_r, $\psi D_{p\ell}$, $N_{p\ell}$ are elements of the generalized Bezout identity (28). Moreover,

37 $\det Y(\zeta) \neq 0 \quad \forall\zeta \in Z[\psi]$,

whence

38 $(\psi X, Y)$ is ℓ.c. $\Leftrightarrow$ (X, Y) is ℓ.c. $\Leftrightarrow$ (D_k, N_k) is ℓ.c. ∎

39 Comments. (a) Except for (37)-(38), the result is similar to that of Theorem 6.2.39 on the polynomial solutions of the comp. eqn. (6.2.32).
(b) Given a solution (X, Y), we set $(\psi X, Y) =: (D_{c\ell}, N_{c\ell})$ i.e., we obtain the compensator $C = D_{c\ell}^{-1}N_{c\ell} = (\psi X)^{-1}Y$ provided that $\det X \neq 0$. The additional

requirement that $(\psi X, Y)$ be an int. pr. ℓ.f. with ψX row-reduced is handled below.

(c) Condition (37) on $N_{c\ell} := Y$ is essential for tracking (see Comment 7.2.90(c)).

42 Proof of Theorem 33. The proof of (a) the equivalence between (34) and (35)-(36) and (b) (X, Y) is ℓ.c. $\Leftrightarrow$ (D_k, N_k) is ℓ.c. uses the generalized Bezout identity (28) and repeats the steps of the proof of Theorem 6.2.39. Condition (37) holds because (a) in (28) $\det U_r(\zeta) \neq 0 \quad \forall \zeta \in Z[\psi]$ (exercise), (b) $\det D_k(\zeta) \neq 0 \quad \forall \zeta \in Z[\psi] \subset \mathbb{C}_+$, and (c) by (36)

$$\det Y(\zeta) = \det D_k(\zeta) \det U_r(\zeta) \neq 0 \quad \forall \zeta \in Z[\psi].$$

Finally by condition (37), $(\psi X, Y)$ is ℓ.c. $\Leftrightarrow$ (X, Y) is ℓ.c. Hence we are done. ∎

We now give conditions under which the solutions (X, Y) of the tracking comp. eqn. (18) are s.t. $(\psi X)^{-1}Y$ is an int. proper left fraction. The proof of the result below is similar to the proof of Theorem 6.2.61 and is therefore omitted.

45 Theorem [Internally proper solutions of tracking comp. eqn. (18)]. Consider the tracking comp. eqn. (18), where (a) $P = N_{pr} D_{pr}^{-1}$ satisfies (4)-(7), (b) $\psi \in \mathbb{R}[s]$ is a monic polynomial s.t. (3) holds, and (c) $D_k \in \mathbb{R}[s]^{n\times n}$ s.t. (16)-(17) holds.

U.t.c.

The tracking compensator equation, given by

18 $$X\psi D_{pr} + YN_{pr} = D_k,$$

has a solution (X, Y) s.t.

19 $$(X, Y) \in \mathbb{R}[s]^{n\times n} \times \mathbb{R}[s]^{n\times n}$$

and

20 the left fraction $(\psi X, Y)$ is int. proper with ψX row reduced with row degrees r_i, $i \in \underline{n}$, and highest degree coefficient matrix X_h

if and only if

46 (a) D_k is row-column-reduced with row powers r_i, $i \in \underline{n}$, resp. column powers k_j, $j \in \underline{n}$,

or equiv.,

47 $D_k(s) = \text{diag}[s^{r_i}]_{i=1}^{n}\ D_{k-}(s)\ \text{diag}[s^{k_j}]_{j=1}^{n}$, where $D_{k-} \in \mathbb{R}(s)^{n\times n}$ is biproper and the highest degree coefficient matrix D_{kh} of D_k satisfies

48 $$D_{kh} = D_{k-}(\infty);$$

49 (b) for the given D_k, (X, Y) is a polynomial matrix solution of tracking comp. eqn. (18) s.t. the row degrees of Y satisfy

$$\partial_{ri}[Y] \leq r_i \quad \forall i \in \underline{n}.$$

Moreover, under conditions (a)-(b) we have that the highest degree coefficient matrices of D_{pr}, ψX, and D_k are related by

50 $$D_{kh} = X_h\ D_{ph}$$

s.t., if $D_{ph} = I$ and $D_{kh} = I$, then $X_h = I$. ∎

51 Comments. (a) In (20), note that ψX is row reduced iff X is row reduced; moreover, with $\psi \in \mathbb{R}[s]$ a monic polynomial the highest row degree coefficient matrices of ψX and X are identical.

(b) From Comment 39(b) and the expanded statement (20), the order of the solving compensator C is $\sum_{i=1}^{n} r_i$, where the r_i's are the row powers of the chosen matrix D_k (see (46)). We shall see below that if these row powers r_i are sufficiently large, then a solution (X, Y) satisfying (49) will exist. Hence again we must choose $\chi(s) = \det D_k(s)$, the char. poly. of Σ, to be of sufficiently large degree for problem TRC to have a solution. ∎

A reasoning similar to that of Theorem 6.2.84 now leads to:

55 Theorem [Existence of int. pr. solutions of tracking comp. eqn. (18)]. Consider the tracking comp. eqn. (18) under the assumptions of Theorem 45.

56 Let $(D_{p\ell}, N_{p\ell})$ be any int. pr. ℓ.c.f. of P with $D_{p\ell}$ row reduced. Then the tracking comp. eqn. (18) has a solution (X, Y) s.t. (19)-(20) hold

if

46 $D_k \in \mathbb{R}[s]^{n\times n}$ is r.c.r. with row powers r_i, $i \in \underline{n}$, and column powers k_j, $j \in \underline{n}$,

s.t.

57 $$r_i \geq \partial\psi + \mu - 1 \quad \forall i \in \underline{n},$$

where

58 $$\mu := \max\{\partial_{ri}[D_{p\ell}],\ i \in \underline{n}\},$$

or equiv.,

59 μ is the maximal degree of any entry of $D_{p\ell}$ where $D_{p\ell}$ is the left plant denominator in (56). ∎

Proof. Observe that in (36) Y may be viewed as the remainder of the division of $D_k U_r$ on the right by $\psi D_{p\ell}$ and repeat the reasoning of the proof of Theorem 6.2.84. ∎

Chapter 8. Design with Stable Plants

8.1. Introduction

In this chapter, Sect. 2 develops the theory of unity feedback systems with stable plants. Section 3 uses this theory to achieve decoupling and eigenvalue placement. Section 4 shows how in the case of an unstable plant one may first stabilize it by local feedback.

In Sect. 2, a simple frequency domain analysis (8.2.6) displays a parameter matrix Q, in terms of which all closed-loop transfer functions are easily calculated. Furthermore, for a given exponentially stable plant, all proper and exp. stable Q's globally parametrize all proper compensators which result in an exp. stable unity feedback system (Theorem 8.2.30). Thus the main design task of providing an exp. stable feedback system and a proper compensator is automatically satisfied. We are left with the design task of meeting other specifications: e.g., achieving a small sensitivity over a given band without saturating the plant. There may be some other limitations on the achievable I/O maps: e.g., Theorem 8.2.45 proves that any plant-$\mathbb{C}_+$-zero is a $\mathbb{C}_+$-zero of the I/O map. Furthermore, Theorem 8.2.70 displays a lower bound on desensitization due to plant-$\mathbb{C}_+$-zeros.

Section 3 contains, for a given exp. stable plant, a Q-parameter algorithm for the design of a compensator, such that the resulting unity feedback system achieves a decoupled (equiv. diagonal) I/O map. This includes the assignment of closed-loop eigenvalues channel by channel.

Section 4 shows how the advantages of Q-parametrization can be extended to the case of unstable plants. Here an unstable plant is first locally stabilized, and thereafter a global parametrization of all exp. stable two-loop feedback systems is found similar to Theorem 8.2.30 (Th. 8.4.18). We obtain finally that all I/O maps, achievable by a single-loop exp. stable feedback system, can be achieved by the two-step scheme described above (Exercise 8.4.24).

8.2. Q-Parametrization Design Properties [Zam. 1, Des. 5]

In this chapter we shall consider the feedback system Σ, shown in Fig. 1, where P is the plant, C is the compensator, and where u_1, u_2, e_1, e_2, y_1, y_2

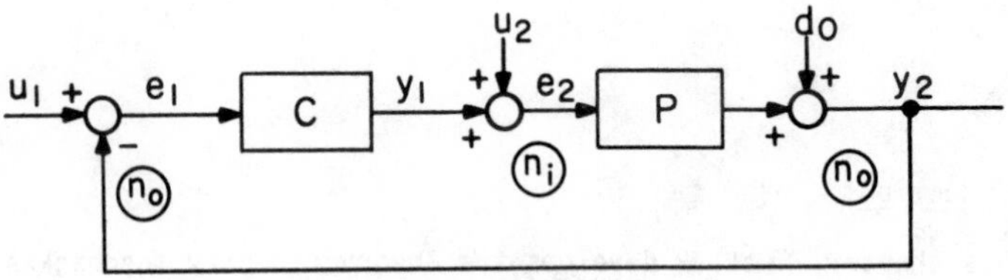

Fig. 1. The feedback system Σ under consideration.

are the usual inputs and outputs; moreover, d_o is a possible disturbance at the output of the plant

1 Assumptions. For feedback system Σ, it is assumed that

2 $P \in \mathbb{R}_{p,o}(s)^{n_o \times n_i}$

3^a $C \in \mathbb{R}(s)^{n_i \times n_o}$; furthermore, using P and C we can define the parameter $Q \in \mathbb{R}(s)^{n_i \times n_o}$ by

3^b $Q := C(I + PC)^{-1} = (I + CP)^{-1}C$;

we assume that

3^c $Q \in \mathbb{R}_p(s)^{n_i \times n_o}$. ∎

In Theorem 14, we shall show that (3^c) will guarantee that all closed-loop transfer functions of Σ are proper.

6 Analysis. As soon as (2) and (3^a) hold and $(I + PC)^{-1} \in E(\mathbb{R}(s))$, we can perform the following algebra:

$$I - PQ = (I + PC)^{-1}, \tag{7}$$

$$I - QP = (I + CP)^{-1}. \tag{8}$$

Hence the global input-output and input-error transfer functions of Σ, viz., H_{yu} resp. H_{eu}, have the following form:

9a $$H_{yu} = \begin{bmatrix} H_{y_1u_1} & H_{y_1u_2} \\ H_{y_2u_1} & H_{y_2u_2} \end{bmatrix} = \begin{bmatrix} C(I+PC)^{-1} & -CP(I+CP)^{-1} \\ PC(I+PC)^{-1} & P(I+CP)^{-1} \end{bmatrix}$$

9b $$= \begin{bmatrix} Q & -QP \\ PQ & P(I-QP) \end{bmatrix},$$

10a $$H_{eu} = \begin{bmatrix} H_{e_1u_1} & H_{e_1u_2} \\ H_{e_2u_1} & H_{e_2u_2} \end{bmatrix} = \left[\begin{array}{c|c} (I+PC)^{-1} & -P(I+CP)^{-1} \\ \hline C(I+PC)^{-1} & (I+CP)^{-1} \end{array}\right]$$

10b $$= \left[\begin{array}{c|c} I-PQ & -P(I-QP) \\ \hline Q & I-QP \end{array}\right].$$

Moreover, H_{yu} and H_{eu} are related by

11a $$H_{eu} = I + FH_{yu}$$

11b $$H_{yu} = F - FH_{eu},$$

where

11c $$F = \begin{bmatrix} 0 & -I \\ I & 0 \end{bmatrix}; \text{ hence } F^{-1} = -F.$$

For design purposes note the following relations:

12a $$H_{y_2d_o} = -H_{e_1d_o} = H_{e_1u_1} = (I+PC)^{-1} = I - PQ,$$

12b $$-H_{y_1d_o} = -H_{e_2d_o} = H_{e_2u_1} = C(I+PC)^{-1} = Q,$$

13 $$C = Q(I-PQ)^{-1} = (I-QP)^{-1}Q.$$

Equations (9^b), (10^b), and (13) show that <u>Q together with P determines all the properties of Σ</u>. ■

We now have

14 Theorem [Well posedness]. Consider feedback system Σ, shown in Fig. 1, where we assume that

2 $\quad P \in \mathbb{R}_{p,o}(s)^{n_o \times n_i}$

15 $\quad C \in \mathbb{R}(s)^{n_i \times n_o}$

and

4 $\quad Q := C(I + PC)^{-1} \in \mathbb{R}(s)^{n_i \times n_o}.$

U.t.c.

16 $\quad$ (a) $Q \in E(\mathbb{R}_p(s)) \Leftrightarrow H_{yu}$ and $H_{eu} \in E(\mathbb{R}_p(s))$

17 $\quad$ (b) $Q \in E(\mathbb{R}_p(s)) \Leftrightarrow C \in E(\mathbb{R}_p(s))$

18 $\quad$ (c) $Q \in E(\mathbb{R}_{p,o}(s)) \Leftrightarrow C \in E(\mathbb{R}_{p,o}(s))$.

19 Comments. (a) By eqn. (9), (10), and (12), H_{eu} and $H_{yu} \in E(\mathbb{R}_p(s))$ if and only if all closed-loop transfer functions of feedback system Σ are proper; hence we define feedback system Σ to be well-posed iff all closed-loop transfer functions are well defined and proper. Consequently by (16), once the plant is strictly proper, (see (2)), assumption (3^c) is a necessary and sufficient condition for the well-posedness of Σ.
(b) From (17) it follows also that, with a strictly proper plant, assumption (3) is necessary and sufficient to guarantee a proper compensator: a usual requirement in view of the presence of noise.
(c) From (18) it follows that, with a strictly proper plant, a strictly proper compensator is guaranteed iff Q is strictly proper.

20 Proof of Theorem 14. $\mathbb{R}_p(s)$ is a ring. Hence, since $P \in E(\mathbb{R}_{p,o}(s))$ by assumption (2), it follows from (9)-(10) that (16) holds. Furthermore, since $P \in E(\mathbb{R}_{p,o}(s))$, it follows that (17) and (18) are consequences of (3^b) and (13). ∎

Consider now $R(0)$ the Euclidean ring of exp. stable transfer functions (see Fact 2.2.6), given by

25 $$\mathcal{R}(0) := \{f \in \mathbb{R}_p(s) : f \text{ is analytic in } \mathbb{C}_+\}$$

and let $\mathcal{R}_o(o)$ denote the <u>subring of strictly proper exp. stable transfer functions</u> given by

26 $$\mathcal{R}_o(0) := \{f \in \mathbb{R}_{p,o}(s) : f \text{ is analytic in } \mathbb{C}_+\}.$$

27 <u>Discussion of exponential stability</u>. Under assumptions (2)-(3), feedback system Σ, shown in Fig. 1, is an interconnected system as in Sec. 4.2, satisfying Assumptions IS (see (4.2.3)) and WP (see (4.2.19)-(4.2.21)). (Note that assumption WP holds by Theorem 14.) Let us now assume throughout this chapter that <u>the plant P and compensator C have underlying PMDs which are well formed and have no unstable hidden modes as in Assumption PMD</u> (see (4.2.25). Then, in view of Theorem 4.2.84, the following definition makes sense.

28 <u>Definition</u>. Under assumptions (2)-(3), feedback system Σ, shown in Fig. 1, is said to be <u>exp. stable</u> iff the global input-output transfer function $H_{yu} \in \mathcal{E}(\mathcal{R}(0))$.

29 <u>Comment</u>. Note that in view of (11)-(12), Σ is exp. stable iff <u>all</u> closed-loop transfer functions of $\Sigma \in \mathcal{E}(\mathcal{R}(0))$.

We have now

30 <u>Theorem</u> [Global Q-parametrization of all stable feedback systems Σ]. Consider feedback system Σ, shown in Fig. 1, under assumptions (2)-(3). U.t.c., <u>if</u>

31 $$P \in \mathcal{E}(\mathcal{R}_o(0)),$$

<u>then</u>

32 (a) $Q \in \mathcal{E}(\mathcal{R}(0)) \Leftrightarrow \Sigma$ is exp. stable and $C \in \mathcal{E}(\mathbb{R}_p(s))$

33 (b) $Q \in \mathcal{E}(\mathcal{R}_o(0)) \Leftrightarrow \Sigma$ is exp. stable and $C \in \mathcal{E}(\mathbb{R}_{p,o}(s))$. ∎

34 <u>Comment</u>. Theorem 30 teaches us that, given an exp. stable and strictly proper plant P, an exp. stable (and strictly proper) parameter transfer function Q parametrizes <u>all</u> exp. stable feedback systems Σ with a proper

(resp. strictly proper) compensator C. Moreover, eqns. (9)-(13) show that Q determines everything. The design task is therefore reduced to choosing an appropriate $Q \in \mathcal{E}(\mathcal{R}(0))$ (resp. $Q \in \mathcal{E}(\mathcal{R}_o(0))$) that satisfies the design specifications required in addition to closed-loop stability and properness of C (see below).

Proof of Theorem 30. Note that $\mathcal{R}(0)$ is a ring and that Σ is exp. stable iff $H_{yu} \in \mathcal{E}(\mathcal{R}(0))$ (by Definition 28). Note also that $\mathcal{R}_o(0)$ is a subring of $\mathcal{R}(0)$. Hence in view of (31), (32) is a consequence of (9^b), the closure properties of $\mathcal{R}(0)$, and Theorem 14. A similar argument establishes (33). ■

35 Q-parametrization design trade-off. Consider feedback system Σ, shown in Fig. 1, and assume that $P \in \mathcal{E}(R_o(0))$ and $Q \in \mathcal{E}(\mathcal{R}(0))$ (whence, by Theorem 30, Σ is exp. stable and $C = Q(I - PQ)^{-1} \in \mathcal{E}(\mathbb{R}_p(s))$). Then from (9)-(10) we have that

(a) the I/O map $H_{y_2u_1}$ of Σ satisfies

36 $$H_{y_2u_1} = PQ;$$

(b) the input to plant-input map $H_{e_2u_1}$ of Σ satisfies

37 $$H_{e_2u_1} = Q;$$

(c) the sensitivity transfer function of Σ (see Sec. 1.3) satisfies

38 $$(I + PC)^{-1} = H_{e_1u_1} = H_{y_2d_o} = I - PQ.$$

Formulas (36)-(38) are important because they display the basic design trade-off:

(1) From (36), (38), and (1.3.52) we see that if

39 $$\|(I - PQ)(j\omega)\| \ll 1 \quad \forall\omega \in \Omega,$$

where Ω is the frequency band of interest, then, over Ω, (a) the I/O map $H_{y_2u_1}$ is approximately I, (b) we obtain desensitization of the closed-loop system output y_2 due to disturbances d_o applied at the output of the plant, and (c) we obtain desensitization of the I/O map $H_{y_2u_1}$ due to plant variations.

(2) On the other hand, by (37), to prevent saturation at the input of the plant one must also have

40 $$\|Q(j\omega)\| \leq K_1 \quad \forall\omega \in \Omega$$

for some given $K_1 > 0$. ∎

Although (39)-(40) exhibit the main design trade-off, some other limitations may be present due to the fact that the plant has $\mathbb{C}_+$-zeros.

45 Theorem. [Limitations on the I/O map]. Consider feedback system Σ, shown in Fig. 1, where $P \in E(R_o(0))$. Then any achievable I/O map $H_{y_2u_1}$ of an exp. stable feedback system Σ with a proper compensator C is of the form

46 $$H_{y_2u_1} = PQ, \quad Q \in E(R(0)).$$

Hence

47 $$Z[P] \cap \mathbb{C}_+ \subset Z[H_{y_2u_1}] \cap \mathbb{C}_+.$$

48 Comment. Note that due to (46) every $\mathbb{C}_+$-plant zero is a $\mathbb{C}_+$-zero of the I/O map. Note also that due to (46), since $H_{y_2u_1}$ and P are strictly proper and Q is proper, "as $s \to \infty$, $H_{y_2u_1}$ tends to zero at least as fast as P."

50 Proof of Theorem 45. That any achievable I/O map is the form (46) is a consequence of Theorem 30 (exercise). In order to prove (47), let $(D_{p\ell}, N_{p\ell})$ be a ℓ.c.f. of P and let (N_{qr}, D_{qr}) be a r.c.f. of Q. Observe now that, since $P \in E(R_o(0))$ and $Q \in E(R(0))$, $\det[D_{p\ell}(s)] \neq 0$ $\forall s \in \mathbb{C}_+$, and $\det[D_{qr}(s)] \neq 0$ $\forall s \in \mathbb{C}_+$. Hence $H_{y_2u_1} = PQ = D_p^{-1}N_p\, N_{qr}D_{qr}^{-1}$ is a left-right fraction of $H_{y_2u_1}$ with no unstable hidden modes. As a consequence (exercise),
$z \in Z[P] \cap \mathbb{C}_+ \;\Rightarrow\; z \in Z[H_{y_2u_1}] \cap \mathbb{C}_+$. ∎

51 Exercise. For the system Σ of Fig. 1, let (2) and (3) hold and let Σ be exp. stable. Let $n_i \geq n_o$ and normal rank P = normal rank C = n_o. Using int. pr. coprime fractions, we have

$$H_{y_2u_1} = N_{pr}[D_{c\ell}D_{pr} + N_{c\ell}N_{pr}]^{-1}N_{c\ell}.$$

Prove the following statements:

(i) If $z \in \mathbb{C}_+$, then $z \in Z[P] \Rightarrow z \in Z[H_{y_2u_1}]$ and $z \in Z[C] \Rightarrow z \in Z[H_{y_2u_1}]$.

(ii) $\forall s \in \mathbb{C}_+$, $R[N_{pr}(s)] \supset R[H_{y_2u_1}(s)]$. Give a dynamical interpretation when $z \in Z[P] \cap \mathbb{C}_+$.

(iii) $\forall z$ s.t. $\det[D_{c\ell}D_{pr} + N_{c\ell}N_{pr}](z) \neq 0$, $z \in Z[P] \Rightarrow z \in Z[H_{y_2u_1}]$.

52 <u>Exercise</u>. For the system Σ of Fig. 1, let (2) and (3) hold, where P and C are as in Exercise 51; furthermore, insert in the feedback path a transfer function $F(s) \in \mathbb{R}_p(s)^{n_o \times n_o}$ with int. pr. r.c.f. (N_{fr}, D_{fr}). Assume that the resulting feedback system is exp. stable. Prove the following statements:

(i) The I/O map $u_1 \mapsto y_2$ is described by the PMD

$$\mathcal{D}_{y_2u_1} = \left(\begin{bmatrix} D_{c\ell}D_{pr} & N_{c\ell}N_{fr} \\ N_{pr} & -D_{fr} \end{bmatrix}, \begin{bmatrix} N_{c\ell} \\ 0 \end{bmatrix}, [N_{pr} \;\vdots\; 0], 0 \right).$$

(ii) If $z \in \mathbb{C}_+$, $z \in Z[P] \Rightarrow z \in Z[H_{y_2u_1}]$.

(iii) $\forall s \in \mathbb{C}_+$, $R[N_{pr}(s)] \supset R[H_{y_2u_1}(s)]$.

53 <u>Exercise</u>. Consider the system Σ_2 shown in Fig. 2. The plant $P \in \mathbb{R}_{p,o}(s)^{n_o \times n_i}$ has an int. pr. r.c.f. (N_{pr}, D_{pr}). The two-input one-output compensator, is specified by two transfer functions and their int. pr. left fractions

$$G_{y_1v_1} = D_{c\ell}^{-1}N_\pi \quad \text{and} \quad G_{y_1e_1} = D_{c\ell}^{-1}N_{c\ell},$$

where

$$\text{rk}[D_{c\ell} \;\vdots\; N_{c\ell} \;\vdots\; N_\pi](s) = n_o \quad \forall s \in \mathbb{C}.$$

Assume that Σ_2 is exp. stable. Prove the following.

(i) Write $D_k := D_{c\ell}D_{pr} + N_{c\ell}N_{pr}$; then

$$H_{y_2u_1} = N_{pr}D_k^{-1}N_\pi.$$

(ii) $\forall z \in \mathbb{C}_+$, $z \in Z[P] \Rightarrow z \in Z[H_{y_2u_1}]$.

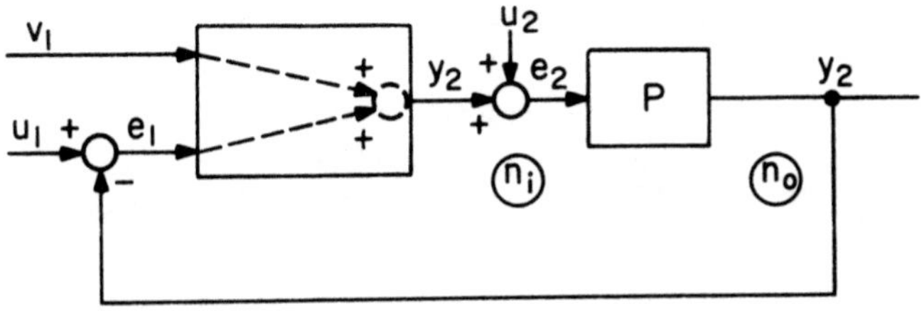

Fig. 2. System Σ_2.

(iii) $\forall s \in \mathbb{C}_+$, $\mathcal{R}[N_{pr}(s)] \supset \mathcal{R}[H_{y_2u_1}(s)]$.

In (4.4.3) we have seen for system Σ of Fig. 1 that, if P has $\mathbb{C}_+$-zeros and if the loop gain is multiplied by some factor $k \nearrow \infty$, then each $\mathbb{C}_+$-zero of the plant is approached arbitrarily closely by a closed-loop eigenvalue. In view of Sec. 1.3 this suggests that the achievable desensitization in (39) is bounded because of these $\mathbb{C}_+$-zeros. We propose to consider this problem from a slightly different point of view.

Bound on achievable desensitization due to plant $\mathbb{C}_+$-zeros. Let the plant be specified by

60 $$P \in \mathcal{R}_o(0)^{n_o \times n_i} \text{ and normal rank of } P = n_o .$$

Consider the system Σ, shown in Fig. 1, with

61 $$Q \in \mathcal{R}_o(0)^{n_i \times n_o} .$$

Hence by Theorem 30, Σ is exp. stable. Note now that Fig. 1 is Fig. 1.3.1 with F = I: the sensitivity transfer function of Σ is (using $Q := C(I + PC)^{-1}$)

38 $$(I + PC)^{-1} = H_{e_1u_1} = H_{y_2d_o} = I - PQ,$$

and, from eqn. (1.3.32), if P is slightly perturbed to $\tilde{P}$, then, with higher-order terms in $(\tilde{P} - P)$ neglected,

62 $$\Delta H_{y_2u_1} = (I + PC)^{-1} \Delta H^o_{y_2u_1} = (I - PQ) \Delta H^o_{y_2u_1} .$$

Thus to achieve desensitization we should keep $I - PQ$ as small as possible over the frequency band Ω of interest (see (39)).

63 Remark. Any realistic model requires a strictly proper P and at least a proper C (hence $Q = C(I + PC)^{-1}$ is proper); consequently, as $\omega \to \infty$ $(I - PQ)(j\omega) \to I$. Hence we can keep $I - PQ$ small only over some finite band Ω of frequencies.

Following [Zam.1] we use a weighting function

64 $$w(s) \in \mathcal{R}_o(0).$$

Typically, $|w(j\omega)| \simeq 1$, $\forall \omega \in \Omega$, where Ω is the band of interest, and drops sharply to zero outside (e.g., Butterworth functions, [Tem.1]). The design must be closed-loop exp. stable ($\Leftrightarrow Q \in \mathcal{E}(\mathcal{R}(0))$) and must give the largest amount of desensitization: hence the minimum problem

65 $$\inf\{\sup_{\omega \geq 0} \|w(j\omega)(I - PQ)(j\omega)\|_\infty : Q \in \mathcal{R}(0)^{n_i \times n_o}\} =: m(P).$$

We use ℓ_∞-induced norms for technical reasons. Note that for $Q = 0$, $m(P) = 1$.

Let the plant P in (60) have $z_1, z_2, \cdots, z_k$ as $\mathbb{C}_+$-zeros: hence

66 $$\text{rk}[P(z_i)] < n_o \quad \forall i \in \underline{k}$$

with $\text{rk}[P(z)] = n_o$ elsewhere in $\mathbb{C}_+$.

A lower bound on achievable desensitization is given by the following.

70 Theorem [Lower bound on desensitization]. For the plant P satisfying (60) and (66), with the weighting function subject to (64), the minimum of problem (65) satisfies

71 $$m(P) \geq |w(z_i)|, \quad \forall i \in \underline{k}.$$

72 Comment. The lower bound in (71) depends on the location of the z_i's in $\mathbb{C}_+$. For example, for $w(s) = (1+s)^{-n}$, if z_i is definitely outside the band of interest (e.g., $|z_i| \gg 1$), the corresponding $|w(z_i)|$ is very small. However, if $|z_i| < 1$, $|w(z_i)|$ may be close to 1.

73 Proof of Theorem 70. $\forall i \in \underline{k}$, $\forall Q \in \mathcal{R}(0)^{n_i \times n_o}$, $P(z_i)\,Q(z_i)$ is singular; hence choose $\eta_i \in \mathbb{C}^{n_o}$ with $\|\eta_i\| = 1$ (here $\|\cdot\|$ denotes ℓ_∞-norms), such that

74 $$P(z_i)Q(z_i)\eta_i = \theta.$$

Let

75 $$f(s, \eta) := w(s)[I - P(s)Q(s)]\eta$$

with $f : \mathbb{C}_+ \times \mathbb{C}^{n_o} \to \mathbb{C}^{n_o}$; note that, $\forall \eta$, $s \mapsto f(s,\eta)$ is analytic in $\mathbb{C}_+$. Now, by (65),

76 $$m(P) = \inf_Q \sup_\omega \sup_\eta \|f(j\omega, \eta)\|,$$

where $Q \in \mathcal{R}(0)^{n_i \times n_o}$, $\omega \geq 0$, $\|\eta\| = 1$. Hence

77 $$m(P) \geq \inf_Q \sup_i \sup_\omega \|f(j\omega, \eta_i)\|.$$

By the maximum modulus theorem and (74), we have considering the ℓth component:

78 $$\sup_\omega |f_\ell(j\omega, \eta_i)| \geq |f_\ell(z_i, \eta_i)| = |w(z_i)\,\eta_{i\ell}|,$$

where the subscript ℓ denotes the ℓth component of the corresponding vector. Now, by (78) and $\|\eta_i\| = 1$,

79 $$\sup_\omega \|f(j\omega, \eta_i)\| \geq \max_\ell |w(z_i)\,\eta_{i\ell}| = |w(z_i)|.$$

Since the dependence on Q has disappeared from (79), (77) becomes

80 $$m(P) \geq \sup_i |w(z_i)|.$$ ■

8.3. Q-Design Algorithm for Decoupling by Feedback

To illustrate the power of the results of Sec. 8.2, we shall consider a plant $P \in \mathcal{R}_o(0)^{n \times n}$ and display a procedure for obtaining a strictly proper compensator C s.t.

(i) the closed-loop feedback system Σ of Fig. 8.2.1 is exp. stable;

(ii) the I/O map $H_{y_2u_1}$ is decoupled and strictly proper;

(iii) in each diagonal element, the poles and zeros (in addition to the $\mathbb{C}_+$-zeros of P) can be specified by the designer.

1 Algorithm [Decoupled I/O map] [Des. 5].

Data: $P \in R_o(0)^{n\times n}$.

Step 1. Obtain an int. pr. r.c.f. (N_{pr}, D_{pr}) of P.

Step 2. If $Z[P] \cap \mathbb{C}_+ = \phi$, pick $n_{j+}(s) \equiv 1 \quad \forall j \in \underline{n}$, and go to step 4; otherwise, go to step 3.

Step 3. Calculate $[\gamma_{ij}]_{i,j\in\underline{n}} =: N_{pr}^{-1}$

and

choose n polynomials $n_{j+}(s)$ of least degree s.t.

2 $\forall i \in \underline{n} \quad \gamma_{ij}(s)n_{j+}(s)$ is analytic in $\mathbb{C}_+$.

Comment: For each j, the polynomial n_{j+} will cancel all the $\mathbb{C}_+$-poles of the jth column of N_{pr}^{-1}. Hence $P^{-1}\cdot\text{diag}[n_{j+}]_{j=1}^{n}$ will be analytic in $\mathbb{C}_+$: we have canceled all $\mathbb{C}_+$-poles of P^{-1}.

Step 4. Choose $\tilde{n}_j(s)$ and $d_j(s)$ $\forall j \in \underline{n}$ in

3 $$H_{y_2u_1}(s) := \text{diag}\left[\frac{n_{j+}(s)\tilde{n}_j(s)}{d_j(s)}\right]_{j=1}^{n}$$

s.t. for every j

4 (i) $Z[d_j] \subset \mathring{\mathbb{C}}_-$,

5 (ii) the polynomial $\tilde{n}_j$ is chosen freely,

6 (iii) $\partial[d_j] > \partial[n_{j+}] + \partial[\tilde{n}_j] + \partial\gamma_j[P^{-1}]$,

where $\partial\gamma_j[P^{-1}]$ denotes the largest degree difference between the numerator and denominator of any element of column j of P^{-1}.

Comments. (a) From (8.2.9), we have $Q = P^{-1}\, H_{y_2u_1}$, whence (2)-(4) guarantee that Q has no $\mathbb{C}_+$-poles, and (6) guarantees that Q is strictly proper. As a consequence, $Q \in E(R_0(0))$ ($\Leftrightarrow \Sigma$, shown in Fig. 8.2.1 with $P \in E(R_0(0))$, is exp. stable (Theorem 8.2.30)).

(b) For obtaining $\partial\gamma_j[P^{-1}]$ the computation of $P^{-1} = D_{pr} N_{pr}^{-1}$ is needed. (Note that P^{-1} is also needed in step 5 below.

Step 5. Calculate the required controller transfer function: let

7 $$n_j(s) := n_{j+}(s)\, \tilde{n}_j(s) \quad \forall j \in \underline{n};$$

then

8 $$C(s) := P(s)^{-1} \operatorname{diag}\left[\frac{n_j(s)}{d_j(s) - n_j(s)}\right]_{j=1}^{n}.$$

Comment. Equation (8) follows from (8.2.13) and (8.2.9). We have $C = Q[I - PQ]^{-1}$, $Q = P^{-1}H_{y_2u_1}$. From (3) and (7) we have $H_{y_2u_1} = \operatorname{diag}[n_j/d_j]$. Hence $C = P^{-1}H_{y_2u_1}(I - H_{y_2u_1})^{-1} = P^{-1}\operatorname{diag}[n_j/d_j]\operatorname{diag}[d_j/(d_j-n_j)]$ $= P^{-1}\operatorname{diag}[n_j/(d_j-n_j)]$.

9 Example. Consider

$$P(s) := \frac{1}{(s+2)^2(s+3)} \left[\begin{array}{c|c} 3s+8 & 2s^2+6s+2 \\ \hline s^2+6s+2 & 3s^2+7s+8 \end{array}\right] \in R_0(0)^{2\times 2},$$

which has a right coprime factorization

$$P(s) = N_{pr}(s)\, D_{pr}(s)^{-1} = \begin{bmatrix} 3 & 2 \\ s+2 & 3 \end{bmatrix} \begin{bmatrix} s^2+3s+4 & 2 \\ 2 & s+4 \end{bmatrix}^{-1}.$$

Since $Z[P] = Z[\det N_{pr}] = \{2.5\} \subset \mathbb{C}_+$, we must go through step 3. Now since

$$N_{pr}(s)^{-1} = \begin{bmatrix} \dfrac{-1.5}{s-2.5} & \dfrac{1}{s-2.5} \\[2ex] \dfrac{0.5(s+2)}{s-2.5} & \dfrac{-1.5}{s-2.5} \end{bmatrix},$$

we choose $n_{1+}(s) = n_{2+}(s) = s-2.5$.

So using $Q = P^{-1}H_{y_2u_1}$, with $H_{y_2u_1}$ given by (3), we have

$$Q(s) = -0.5 \begin{bmatrix} \dfrac{(3s^2+7s+8)\tilde{n}_1(s)}{d_1(s)} & \dfrac{-(2s^2+6s+2)\tilde{n}_2(s)}{d_2(s)} \\ \dfrac{-(s^2+6s+2)\tilde{n}_1(s)}{d_1(s)} & \dfrac{(3s+8)\tilde{n}_2(s)}{d_2(s)} \end{bmatrix}.$$

To guarantee $Q \in \mathcal{R}_o(0)^{2\times 2}$, we choose $\tilde{n}_1(s) = \tilde{n}_2(s) = 1$ and $d_1(s)$, $d_2(s)$ s.t.

(i) $\partial[d_1] \geq 3$, $\partial[d_2] \geq 3$,

(ii) $Z[d_j] \subset \mathring{\mathbb{C}}_-$ for $j = 1, 2$.

Then the resulting I/O map $H_{y_2u_1}$ is given by

10 $$H_{y_2u_1}(s) = \text{diag}\left[\frac{s-2.5}{d_1(s)}, \frac{s-2.5}{d_2(s)}\right],$$

and the resulting compensator is given by

11 $$C(s) = \begin{bmatrix} \dfrac{-0.5(3s^2+7s+8)}{d_1(s)-(s-2.5)} & \dfrac{s^2+3s+1}{d_2(s)-(s-2.5)} \\ \dfrac{0.5(s^2+6s+2)}{d_1(s)-(s-2.5)} & \dfrac{-0.5(3s+8)}{d_2(s)-(s-2.5)} \end{bmatrix}.$$

Note finally that if in (10) $H_{y_2u_1}(s)\Big|_{s=0} = I$, feedback system Σ will asymptotically track any step input. ⌘

12 Exercise. Consider Example 9 and construct a compensator C such that feedback system Σ is exp. stable, tracks asymptotically any ramp, and $H_{y_2u_1}$ is diagonal. (See Exercise 7.2.71.) ⌘

8.4. Two-Step Compensation Theorem for Unstable Plants

In this section we explore the following questions: (i) if the plant P is exp. unstable, may we first stabilize it and then, for the purpose of design, find a global parametrization of all stable feedback systems similar to Theorem 8.2.30? (ii) If so, is any achievable I/O map $H_{y_2u_1}$ achievable by the two-step scheme described in (i)?

Let us first specify our assumptions

1 Assumption. The plant is specified by its transfer function $P_o \in \mathbb{R}_{p,o}(s)^{n_o \times n_i}$, which is generated by a well-formed plant PMD $\mathcal{D}_{P_o}$ with no unstable hidden modes and s.t.

$$P_o \notin E(\mathcal{R}_o(0)).$$

2 Assumption. The plant P_o can be stabilized by an exp. stable local feedback, equiv. there exists a feedback $F_o \in \mathcal{R}(0)^{n_i \times n_o}$, generated by a well-formed PMD $\mathcal{D}_{f_o}$ with no unstable hidden modes, which results in an exp. stable single-loop feedback system ${}^1S(F_o, P_o)$ (see Fig. 1), with the stabilized plant I/O map given by

3
$$H_{y_2'e_2''} =: P_1 = P_o(I + F_oP_o)^{-1} = (I + P_oF_o)^{-1}P_o.$$

4 Comments. (a) The purpose of Assumption 1 is to guarantee the existence of an int. pr. compensator $C \in \mathbb{R}_p(s)^{n_i \times n_o}$ s.t. the single-loop feedback system ${}^1S(P_o, C)$, shown in Fig. 2 is exp. stable.

For example, $P_o \in \mathbb{R}_{p,o}(s)^{n_o \times n_i}$ is generated by the well-formed PMD $\mathcal{D}_{P_o} = [D_{P_or}, I, N_{P_or}, 0]$, where (N_{P_or}, D_{P_or}) is an int. pr. r.c.f. of P_o with D_{P_or} column reduced; by the method of Sec. 6.2, we can then find an int. pr. compensator $C \in \mathbb{R}_p(s)^{n_i \times n_o}$ s.t. ${}^1S(P_o, C)$ is exp. stable.

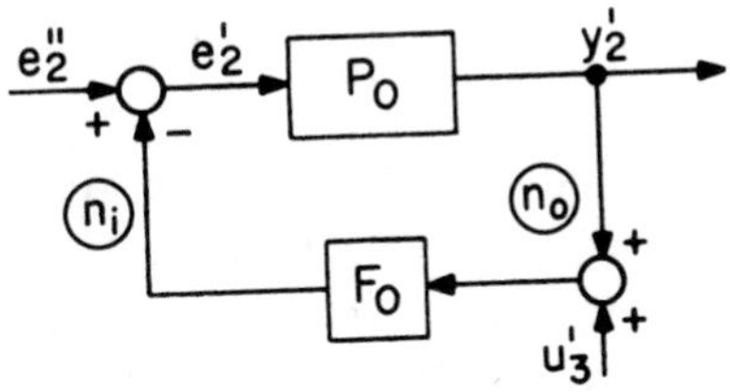

Fig. 1. The system ${}^1S(F_o, P_o)$.

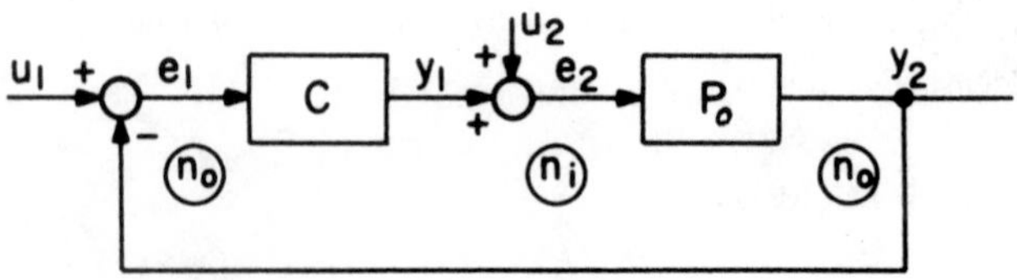

Fig. 2. The system ${}^1S(P_o, C)$.

(b) The notation used in Fig. 1 for system ${}^1S(F_o, P_o)$ will become clear in the sequel.
(c) Comparing Fig. 1 and Fig. 2, we see that systems ${}^1S(F_o, P_o)$ and ${}^1S(P_o, C)$ are similar if we interchange F_o with P_o and P_o with C. Hence in view of definition (8.2.3^b) the parameter of ${}^1S(F_o, P_o)$ reads $P_o(I + F_oP_o)^{-1} = P_1$. Consequently, by an analysis similar to the one done for Theorem 8.2.30, we have

5 Fact. Consider the single-loop feedback system ${}^1S(F_o, P_o)$, shown in Fig. 1, where P_o satisfies Assumption 1 and $F_o \in \mathcal{E}(\mathcal{R}(0))$ is as in Assumption 2.
U.t.c.

6 $${}^1S(F_o, P_o) \text{ is exp. stable}$$

$\Leftrightarrow$

the transformed-plant I/O map $P_1 : e_2'' \mapsto y_2'$ satisfies

7 $$P_1 := P_o(I + F_oP_o)^{-1} \in \mathcal{E}(\mathcal{R}_o(0)).$$ ∎

8 Comments. (a) Note that system ${}^1S(F_o, P_o)$ (see Fig. 1) reduces to P_1 if $u_3' = \theta$ and that by Fact 5 the exp. stability of P_1 is equivalent to the exp. stability of ${}^1S(F_o, P_o)$. Although subsystem P_1 is simpler than system ${}^1S(F_o, P_o)$, the additional external input u_3' provides an additional degree of freedom which enables us to realize all achievable I/O maps $H_{y_2u_1}$ of ${}^1S(P_o, C)$ (see Fig. 2) by the two-loop feedback system studied below (see Exercise 24).

(b) The Assumption 2 that P_o, given by Assumption 1, can be stabilized by an exp. stable feedback $F_0 \in \mathcal{R}(0)^{n_i \times n_o}$ is not very restrictive. Indeed [You.1], this is always possible if there exists no finite point z on the nonnegative real axis in $\mathbb{C}_+$ s.t. $P_o(z) = 0$. If P_o is zero at distinct finite points $z_1, z_2, \cdots, z_\ell$ on the nonnegative real axis in $\mathbb{C}_+$, let ν_i denote the number of $\mathbb{C}_+$-poles of $P_o \in \mathbb{R}_{p,o}(s)^{n_o \times n_i}$ on the nonnegative real axis to the right of z_i, counting multiplicities according to their Mcmillan degrees, (i.e., their maximal order as a pole of any minor of any order of P_o). Then P_o is stabilizable by a stable F_0 if and only if the integers ν_i are all even, [This follows by (a) $P_o(\infty) = 0$ and (b) [You.1, Th. 2, Corr. 1]].

We shall now compensate the exp. stable system ${}^1S(F_0, P_o)$ by $C - F_0$, thus giving the two-loop feedback system ${}^3S(P_o, F_0, C - F_0)$, shown in Fig. 3.

9 Forward Comment. Roughly speaking, in Theorem 18 we shall show (a) that the existence of a proper compensator C resulting in an exp. stable system ${}^1S(P_o, C)$ (Fig. 2) is equivalent to the existence of the proper compensator $C - F_0$ s.t. ${}^3S(P_o, F_0, C - F_0)$ is exp. stable, and (b) that all stable systems ${}^3S$ can be globally parametrized by a parameter $Q \in E(\mathcal{R}(0))$ s.t. $C - F_0 = Q(I - P_1Q)^{-1}$, where P_1 is the stabilized plant (3) (whence by (a) all stable systems ${}^1S(P_o, C)$ can be parametrized by an exp. stable parameter Q as in

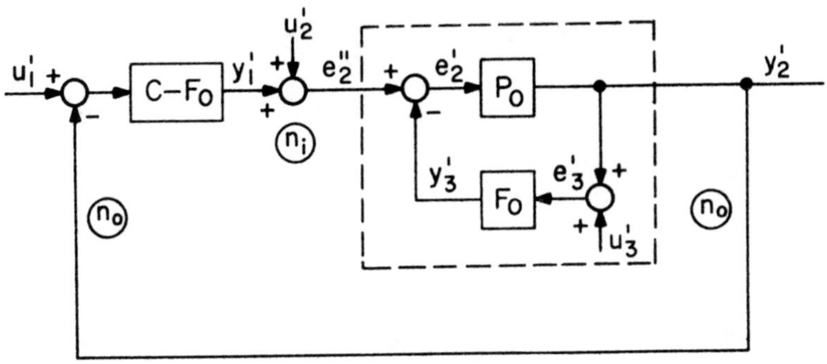

Fig. 3. The system ${}^3S(P_o, F_0, C - F_0)$.

Theorem 8.2.30). Moreover, in Exercise 24 below we shall show that any I/O map $H_{y_2u_1}$, achievable by an exp. stable system ${}^1S(P_o, C)$, can be achieved as the I/O map $H_{y_2'u_1'}$ of the corresponding exp. stable system ${}^3S(P_o, F_0, C - F_0)$ by setting $u_3' = -u_1'$. <u>Consequently, from an algebraic point of view the two-step compensation scheme</u> ${}^3S(P_o, F_0, C - F_0)$ <u>can do whatever the one-step scheme</u> ${}^1S(P_o, C)$ <u>can do</u>. <u>No design opportunity is lost</u>.

10 <u>Analysis</u>. Consider systems ${}^1S(P_o, C)$ and ${}^3S(P_o, F_0, C - F_0)$, shown in Figs. 2 and 3, under Assumptions 1 and 2. Using Laplace transforms, and merging the two middle summing nodes in Fig. 3 into one node, the system equations for ${}^1S$ resp. ${}^3S$ are

$$11 \quad \underbrace{\begin{bmatrix} I & P_o \\ -C & I \end{bmatrix}}_{=:\,M} \begin{bmatrix} e_1 \\ e_2 \end{bmatrix} = \begin{bmatrix} u_1 \\ u_2 \end{bmatrix}, \text{ resp. } \underbrace{\begin{bmatrix} I & P_o & 0 \\ -(C-F_0) & I & F_0 \\ 0 & -P_o & I \end{bmatrix}}_{=:\,M'} \begin{bmatrix} e_1' \\ e_2' \\ e_3' \end{bmatrix} = \begin{bmatrix} u_1' \\ u_2' \\ u_3' \end{bmatrix},$$

$$12 \quad \begin{bmatrix} e_1 \\ e_2 \end{bmatrix} = \begin{bmatrix} u_1 \\ u_2 \end{bmatrix} + \underbrace{\begin{bmatrix} 0 & -I \\ I & 0 \end{bmatrix}}_{=:\,F} \begin{bmatrix} y_1 \\ y_2 \end{bmatrix}, \text{ resp. } \begin{bmatrix} e_1' \\ e_2' \\ e_3' \end{bmatrix} = \begin{bmatrix} u_1' \\ u_2' \\ u_3' \end{bmatrix} + \underbrace{\begin{bmatrix} 0 & -I & 0 \\ I & 0 & -I \\ 0 & I & 0 \end{bmatrix}}_{=:\,F'} \begin{bmatrix} y_1' \\ y_2' \\ y_3' \end{bmatrix}.$$

Let

$$n := n_i + n_o, \qquad \text{resp. } n' := n_i + n_o + n_o.$$

Note that F and F' are <u>constant</u>, that F is <u>nonsingular</u> and $F_0 \in \mathcal{R}(0)^{n_i \times n_o}$; consequently, using (11)-(12) and Fig. 3

13 $^1S(P_o, C)$ is exp. stable $\Leftrightarrow$ $H_{yu} \in R(0)^{n\times n}$ $\Leftrightarrow$ $H_{eu} \in R(0)^{n\times n}$

$$\Leftrightarrow \; M^{-1} \in R(0)^{n\times n},$$

and

14 $^3S(P_o, F_o, C - F_o)$ is exp. stable $\Leftrightarrow$ $H_{y'u'} \in R(0)^{n'\times n'}$

$$\Leftrightarrow \; H_{e'u'} \in R(0)^{n'\times n'} \; \Leftrightarrow \; (M')^{-1} \in R(0)^{n'\times n'}.$$

Note with care that, by (12), $H_{y'u'} \in R(0)^{n'\times n'} \Rightarrow H_{e'u'} \in R(0)^{n'\times n'}$; moreover, the relations $y_2' = u_1' - e_1'$, $y_3' = F_o e_3'$, with $F_o \in R(0)^{n_i\times n_o}$, and $y_1' = e_2' - u_2' + y_3'$ imply that $H_{e'u'} \in R(0)^{n'\times n'} \Rightarrow H_{y'u'} \in R(0)^{n'\times n'}$ (note that we cannot use (12), since F' is singular).

14 Comment. In the stability analysis above it is assumed that C and $C - F_o$ are generated by well-formed PMDs with no unstable hidden modes, as is the case for P_o and F_o (see Assumptions 1 and 2). In other words, although the analysis is made in the frequency domain in algebraic terms, there is everywhere an implied time-domain interpretation. Moreover, the same applies to the results below.

18 Theorem [Two-step compensation theorem]. Consider the systems $^1S(P_o, C)$ and $^3S(P_o, F_o, C - F_o)$, shown in Figs. 2 and 3, and described by (11)-(12). Let Assumptions 1 and 2 hold.
U.t.c., for any $F_o \in R(0)^{n_i\times n_o}$ satisfying (2),

19 $\exists\, C \in \mathbb{R}_p(s)^{n_i\times n_o}$ s.t. $^1S(P_o, C)$ is exp. stable,

$\Leftrightarrow$

20 $\exists\, C - F_o \in \mathbb{R}_p(s)^{n_i\times n_o}$ s.t. $^3S(P_o, F_o, C - F_o)$ is exp. stable,

$\Leftrightarrow$

21 $\exists\, Q \in R(0)^{n_i\times n_o}$ s.t. $C - F_o := Q(I - P_1 Q)^{-1}$. ∎

22 Remark. Recall Assumptions 1 and 2, Fact 5, and Theorem 8.2.30. Then, with the Q defined in (21), it follows that

$$Q \in R(0)^{n_i\times n_o} \;\Leftrightarrow\; {}^1S(P_1, C - F_o) \text{ is exp. stable},$$

where ${}^1S(P_1, C - F_0)$ is obtained from ${}^3S$ by setting $u_3' = \theta$ in Fig. 3.

23 Exercise. Use the Q, defined in (21), and the P_1, defined in (3), to prove that

$$H_{y_1'u_1'} = Q;\ H_{y_2'u_1'} = P_1Q = (I + P_oF_o)^{-1} P_oQ;$$

$$H_{e_1'u_1'} = I - P_1Q = I - (I + P_oF_o)^{-1}P_oQ.$$

24 Exercise [Achievable I/O maps]. In system ${}^3S(P_o, F_0, C - F_0)$, shown in Fig. 3, set $u_3' = -u_1'$ (equivalently feedforward u_1' through a -I gain to u_3'), and show that for the resulting system, $H_{y_2'u_1'} = H_{y_2u_1}$, where the latter is the I/O map of ${}^1S(P_o, C)$.

(Hint: Use (12): set $u_3' = -u_1'$; perform $\rho_3 \leftarrow \rho_3 + \rho_1$; eliminate e_3'; ....)

25 Exercise. Verify that the forward comment (9) is true. Hence the answer to both questions in the beginning of this section is yes.

26 Comment on the exp. stability of F_0. Consider Figs. 2 and 3, the definition of P_1 in (3), Fact 5, and statements (19), (20), and (21) of Theorem 18. It is intuitively clear that exp. stability of ${}^3S(P_o, F_0, C - F_0)$ implies the exp. stability of ${}^1S(P_o, C)$ and ${}^1S(P_1, C - F_0)$ without requiring that F_0 be exp. stable (this fact will be exhibited in the proofs of (20)⇒(19) and (20)⇒(21) below by appropriately restricting ${}^3S$). However the proofs of (19)⇒(20) and (21)⇒(20) require that F_0 be exp. stable; to emphasize that this assumption is indispensable we offer the corollary below.

30 Corollary. Let $P_o \in \mathbb{R}_{p,o}(s)^{n_o \times n_i}$. Let $C \in \mathbb{R}_p(s)^{n_i \times n_o}$ be s.t. ${}^1S(P_o, C)$, defined by (11) and (12), is exp. stable. Let

31 $F = \{F_0 \in \mathbb{R}_p(s)^{n_i \times n_o}$: F_0 has at least one $\mathbb{C}_+$-pole, say p_1, s.t. p_1 is not a pole of C$\}$.

U.t.c.

32 $\forall F_0 \in F$, ${}^3S(P_o, F_0, C - F_0)$ is not exp. stable.

Proof. On eqn. (11) describing ${}^3S(P_o, C_o, C - F_0)$, perform $\gamma_1 \leftarrow \gamma_1 - \gamma_3$ and $\rho_3 \leftarrow \rho_3 + \rho_1$ to obtain

33 $$\begin{bmatrix} I & P_o & 0 \\ -C & I & F_0 \\ 0 & 0 & I \end{bmatrix} \begin{bmatrix} e_1' \\ e_2' \\ e_3'' \end{bmatrix} = \begin{bmatrix} u_1' \\ u_2' \\ u_4' \end{bmatrix},$$

where $e_3'' := e_1' + e_3'$ and $u_4' := u_1' + u_3'$, Easy calculations using (33) give

$$H_{e_1'u_4'} = P_o(I + CP_o)^{-1}F_0,$$

$$H_{e_2'u_4'} = -(I + CP_o)^{-1}F_0.$$

Hence using int. pr. coprime fractions, these relations become

34a $$H_{e_1'u_4'} = N_{p_or}(D_{c\ell}D_{p_or} + N_{c\ell}N_{p_or})^{-1} D_{c\ell} D_{f_o\ell}^{-1} N_{f_o\ell},$$

34b $$H_{e_2'u_4'} = -D_{p_or}(D_{c\ell}D_{p_or} + N_{c\ell}N_{p_or})^{-1} D_{c\ell} D_{f_o\ell}^{-1} N_{f_o\ell}.$$

Now note that, since (N_{p_or}, D_{p_or}) is r.c., by the Bezout identity, there exist polynomial matrices U, V s.t.

$$UN_{p_or} + VD_{p_or} = I,$$

whence (34) implies

35 $$D_{c\ell}^{-1}(D_{c\ell}D_{p_or} + N_{c\ell}N_{p_or})(UH_{e_1'u_4'} - VH_{e_2'u_4'}) = D_{f_o\ell}^{-1} N_{f_o\ell} = F_0.$$

Since $F_0 \in F$, by (31), the RHS of (35) has a pole at $p_1 \in \mathbb{C}_+$, while $\det[D_{c\ell}(p_1)] \neq 0$. Moreover, since by assumption ${}^1S(P_o, C)$ is exp. stable, $\det[(D_{c\ell}D_{p_or} + N_{c\ell}N_{p_or})(p_1)] \neq 0$. Assume now, for the purpose of a contradiction that ${}^3S(P_o, F_0, C - F_0)$ is exp. stable. Then, with $u_4' = u_1' + u_3'$, $H_{e_1'u_4'}$ and $H_{e_2'u_4'} \in E(R(0))$. Hence, with U and $V \in E(\mathbb{R}[s])$, and $D_{c\ell}^{-1}(D_{c\ell}D_{p_or} + N_{c\ell}N_{p_or})$ nonsingular at $p_1 \in \mathbb{C}_+$, the LHS of (35) has no pole at $p_1 \in \mathbb{C}_+$, while the RHS has a pole there. As a consequence we obtain a contradiction; thus, $\forall F_0 \in F$, ${}^3S(P_o, F_0, C - F_0)$ is not exp. stable. ∎

37 Remark. In particular, the proof shows that with $u_4' = u_1' + u_3'$, $H_{e_1'u_4'}$, and/or $H_{e_2'u_4'}$ must have a pole at $p_1 \in \mathbb{C}_+$. Hence if $u_1'(s) = u_3'(s)$

$= k(s - \alpha)^{-1}$, where $k \in \mathbb{C}^{n_o}$ and $\operatorname{Re} \alpha < 0$, then for some k the responses $e_1'(s)$ and/or $e_2'(s)$ have a pole at $p_1 \in \mathbb{C}_+$ (at least one corresponding time function includes a term $O(\exp p_1 t)$ with $p_1 \in \mathbb{C}_+$).

40 Proof of Theorem 18

(19) ⇒ (20). By assumption ${}^1S(P_o, C)$ is exp. stable, whence, by (13), $M^{-1} \in E(R(0))$. We claim that M', defined in eqn. (11), satisfies $(M')^{-1} \in E(R(0))$, whence, by (14), ${}^3S(P_o, F_o, C - F_o)$ is exp. stable.

On M', perform $\gamma_1 \leftarrow \gamma_1 - \gamma_3$, $\rho_3 \leftarrow \rho_3 + \rho_1$, and $\rho_2 \leftarrow \rho_2 - F_o\rho_3$ to obtain

41 $$M_3 = \begin{bmatrix} I & P_o & 0 \\ -C & I & 0 \\ 0 & 0 & I \end{bmatrix}.$$

Since, by assumption, $F_o \in E(R(0))$, $M_3 = LM'R$, where L and R are unimodular matrices with elements in $R(0)$. Consequently, by inspection of (41), $M^{-1} \in E(R(0))$ implies $(M')^{-1} \in E(R(0))$.

(20) ⇒ (19). By assumption, ${}^3S(P_o, F_o, C - F_o)$ is exp. stable. In (11), set $u_3' = -u_1'$ and perform $\gamma_1 \leftarrow \gamma_1 - \gamma_3$ and $\rho_3 \leftarrow \rho_3 + \rho_1$. Then, with $e_3'' := e_1' + e_3'$,

$$\begin{bmatrix} I & P_o & 0 \\ -C & I & F_o \\ 0 & 0 & I \end{bmatrix} \begin{bmatrix} e_1' \\ e_2' \\ e_3'' \end{bmatrix} = \begin{bmatrix} u_1' \\ u_2' \\ \theta \end{bmatrix}.$$

By the third equation $e_3'' = \theta$, whence

44 $$\begin{bmatrix} I & P_o \\ -C & I \end{bmatrix} \begin{bmatrix} e_1' \\ e_2' \end{bmatrix} = \begin{bmatrix} u_1' \\ u_2' \end{bmatrix}.$$

So with $u_3' = -u_1'$, the partial input-error map $(u_1', u_2') \mapsto (e_1', e_2')$ of ${}^3S$ is identical with the input-error map of ${}^1S(P_o, C)$ (see (11) and (12)). Hence, using (13) and (14), the exp. stability of ${}^3S(P_o, F_o, C - F_o)$ implies the exp. stability of ${}^1S(P_o, C)$.

(20) ⇒ (21). If we show that the exp. stability of ${}^3S(P_o, F_o, C - F_o)$ implies the exp. stability of ${}^1S(P_1, C - F_o)$, then by global parametrization Theorem 8.2.30 and Fact 5, assertion (21) follows.

By assumption, for the system ${}^3S(P_o, F_o, C - F_o)$, the maps $(u_1', u_2', u_3') \mapsto (e_1', e_2', e_3')$ and $(u_1', u_2', u_3') \mapsto (y_1', y_2', y_3')$ are exp. stable (equiv. $\in E(R(0))$). Hence with $e_2'' := y_1' + u_2'$ and $u_3' = \theta$ the partial map $(u_1', u_2') \mapsto (e_1', e_2'')$ is exp. stable. Now, refer to Fig. 3 and observe that $e_2'' = y_1' + u_2'$ is an external input to subsystem ${}^1S(F_o, P_o)$, which reduces to $P_1 := P_o(I + F_oP_o)^{-1}$ when $u_3' = \theta$. Hence, with $u_3' = \theta$, the partial map $(u_1', u_2') \mapsto (e_1', e_2'')$ of ${}^3S$ is the input-error map of ${}^1S(P_1, C - F_o)$: since the latter map is exp. stable, it follows that ${}^1S(P_1, C - F_o)$ is exp. stable.

(21) ⇒ (20)

Step 1. $Q \in E(R(0))$ and $C - F_o = Q(I - P_1 Q)^{-1}$ imply that ${}^1S(P_1, C - F_o)$ is exp. stable by Theorem 8.2.30 and Fact 5. Consequently, with $u_3' = \theta$ (see Fig. 3) and with $e_2' = e_2'' - F_o y_2'$, we conclude that the maps

46 $$\left\{ \begin{array}{l} (u_1', u_2') \mapsto (e_1', e_2''),\ (u_1', u_2') \mapsto (y_1', y_2') \text{ and } (u_1', u_2') \mapsto (e_1', e_2') \\ \text{are exp. stable.} \end{array} \right.$$

Since with $u_3' = \theta$ the eqns. relating (u_1', u_2') to (e_1', e_2') in Fig. 3 are

47 $$\underbrace{\begin{bmatrix} I & P_o \\ -(C - F_o) & I + F_oP_o \end{bmatrix}}_{=:\ M_4} \begin{bmatrix} e_1' \\ e_2' \end{bmatrix} = \begin{bmatrix} u_1' \\ u_2' \end{bmatrix},$$

we conclude that $M_4^{-1} \in E(R(0))$.

Step 2. Consider ${}^3S(P_o, F_o, C - F_o)$. Then on (11) perform $\rho_2 \leftarrow \rho_2 - F_o\rho_3$, $\rho_3 \leftarrow \rho_3 + \rho_1$, and $\gamma_1 \leftarrow \gamma_1 - \gamma_3$ to obtain

50 $$\underbrace{\begin{bmatrix} I & P_o & 0 \\ -(C - F_o) & I + F_oP_o & 0 \\ 0 & 0 & I \end{bmatrix}}_{=:\ M_5} \begin{bmatrix} e_1' \\ e_2' \\ e_1'+e_3' \end{bmatrix} = \begin{bmatrix} u_1' \\ u_2' - F_ou_3' \\ u_3' + u_1' \end{bmatrix}.$$

We have used only row operations in the ring $R(0)$, hence

51 $$(M_5)^{-1} \in E(R(0)) \Leftrightarrow (M')^{-1} \in E(R(0)).$$

Step 3. From step 1, (21) ⇒ $M_4^{-1} \in E(R(0))$. By inspection of (47) and (50), $M_4^{-1} \in E(R(0)) \Leftrightarrow M_5^{-1} \in E(R(0))$. Consequently, by (51) and (14), assumption (21) implies that ${}^3S(P_o, F_0, C - F_0)$ is exp. stable. ∎

52 Comment. Note that the proofs of (20) ⇒ (19) and (20) ⇒ (21) did not require that $F_0 \in E(R(0))$.

Epilogue

At the close of this volume it is important to keep in mind the following points:

1 Discrete-time systems. Lack of space prevented us to follow each chapter with an isomorphic chapter of results covering the discrete-time case, [Kuc.1]. It is very useful to keep in mind the following rough correspondences.

Continuous-time case	Discrete-time case
$f : \mathbb{R}_+ \to \mathbb{R}^n$; $f : t \mapsto f(t)$	$f : \mathbb{N} \to \mathbb{R}^n$, $f = (f_i)_{i=0}^{\infty}$
$\hat{f}$ is analytic in some right-half plane: $\{s \mid \text{Re } s > \rho\}$	$\tilde{f}(z) := \sum_0^{\infty} f_k z^{-k}$ is analytic outside the closed disk $\overline{D(0, \rho_1)}$ centered on 0, with radius ρ_1
$\mathbb{C}_+$	$D(0, 1)^c = \{z \mid \|z\| \geq 1\}$
$\mathring{\mathbb{C}}_-$	$D(0, 1)$ = open disk $= \{z \mid \|z\| < 1\}$
$\mathbb{R}_p(s)$ = proper rational fn	$\tilde{f} \in \mathbb{R}_p(z) \Rightarrow \tilde{f}$ is a causal transfer function
$\hat{f} \in \mathcal{R}(0) \Leftrightarrow \hat{f}$ is proper and exp. st. ($\hat{f}$ analytic in $\mathbb{C}_+$ including at ∞).	$\tilde{f} \in \tilde{\mathcal{R}}(0) \Leftrightarrow \tilde{f}$ is causal and exp. stable ($\tilde{f}$ analytic in $D(0, 1)^c$ including at ∞).

2 Distributed-systems. The concepts results and techniques developed in this volume have been extended to a general class of linear time-invariant distributed systems [Cal.1], [Vid.1].

3 Robustness of stability considerations. Our discussion considered exclusively lumped linear time-invariant systems. As engineers it is important to know that it can be shown that small nonlinear perturbations, slow time-variations of the coefficients and small delay effects do not upset the conclusions on stability and tracking [Des.4], [Wil.1], [Vid.2], [Vid.3], [Saf.2].

4 Computer-aided design. The recent surge of developments in computer-aided design, e.g., [Pol.1] suggests the following views:

(a) The first task of theory is to delineate what is possible.

(b) Design algorithms must provide families of conveniently parametrized designs that satisfy gross qualitative features (e.g., stability, properness, etc.).

Then the design objective is expressed by a performance function which is, e.g., to be minimized subject to a number of inequality constraints that reflect important engineering requirements (e.g., upper bound on some transfer functions to guard against saturation, upper bound on $\omega \mapsto \bar{\sigma}[(I + PC)(j\omega)^{-1}]$ over some interval to guarantee some degree of desensitization, etc.). Then the designer uses the CAD facility to search in parameter space for the "best" solution.

5 The field of multivariable feedback control is rapidly moving. This can be seen from the abundant literature in journals such as the IEEE Transactions on Automatic Control, Automatica, and International Journal of Control, e.g., [Per.1], and research reports, e.g., [Fra.1].

Appendix A: Rings and Fields

1 Every engineer has performed some computations

(a) in the following fields: $\mathbb{R}$, $\mathbb{C}$, Q; $\mathbb{R}(s)$, $\mathbb{C}(s)$;

(b) in the following commutative rings:

$\mathbb{Z}$, $\mathbb{R}[s]$, $\mathbb{C}[s]$, $\mathbb{R}_p(s)$, $\mathbb{R}_{p,o}(s)$, $\mathcal{R}(0)$, $\mathcal{R}_o(0)$, $\mathcal{R}_U$; diagonal matrices with elements in a field (e.g., $\mathbb{R}$, $\mathbb{C}$, or $\mathbb{R}(s)$) or in a commutative ring (e.g., $\mathbb{R}[s]$, $\mathbb{C}[s]$);

where in particular

2 $\mathbb{R}_p(s)$, $(\mathbb{R}_{p,o}(s))$:= the ring of proper[§] (strictly proper)[†] rational functions with coefficients in $\mathbb{R}$.

3 $\mathcal{R}(0)$, $(\mathcal{R}_o(0))$:= the subring of elements of $\mathbb{R}_p(s)$, $(\mathbb{R}_{p,o}(s))$, that are analytic in $\mathbb{C}_+$ (i.e., with no poles in $\mathbb{C}_+$)

4 $\mathcal{R}_{\mathcal{U}}$:= the subring of elements of $\mathbb{R}_p(s)$ that are analytic in $\mathcal{U}$: a closed subset of $\mathbb{C}$ symmetric w.r.t. the real axis and which includes $\mathbb{C}_+$.

Elements of $\mathcal{R}(0)$ and $\mathcal{R}_{\mathcal{U}}$ are also called exp. stable resp. $\mathcal{U}$-stable transfer functions.

(c) in the following noncommutative rings:

$$\mathbb{R}^{n\times n},\ \mathbb{C}^{n\times n};\ \mathbb{R}[s]^{n\times n},\ \mathbb{C}[s]^{n\times n};\ \mathbb{R}(s)^{n\times n},\ \mathbb{C}(s)^{n\times n};\ \mathbb{R}_p(s)^{n\times n},$$

$$\mathbb{R}_{p,o}(s)^{n\times n};\ \mathcal{R}(0)^{n\times n},\ \mathcal{R}_o(0)^{n\times n};\ \mathcal{R}_U^{n\times n},$$

where we consider $n \times n$ matrices with elements in $\mathbb{R}$, $\mathbb{C}$, etc. ■

5 Exercise. For each of the rings and fields above, verify that you already know the operations of addition and multiplication. Identify precisely the element 0 and the element 1, the identity under addition and multiplication, resp.

[§] Bounded at infinity;

[†] (zero at infinity).

6 Definition. Any ring and any field is a set of elements together with two binary operations, an addition + and a multiplication • ; an identity element under addition denoted by 0; and an identity element under multiplication denoted by 1, and they obey the following respective axioms:

Axioms

Ring: $(R, +, \cdot; 0, 1)$	Field: $(\mathbb{F}, +, \cdot; 0, 1)$

Addition is

Associative: $(\alpha + \beta) + \gamma = \alpha + (\beta + \gamma) \quad \forall \alpha, \beta, \gamma$

Commutative: $\alpha + \beta = \beta + \alpha \quad \forall \alpha, \beta$

$\exists$ identity 0: $\alpha + 0 = \alpha \quad \forall \alpha$

$\exists$ inverse: $\forall \alpha$, $\exists$ element $(-\alpha) \ni \quad \alpha + (-\alpha) = 0$

Multiplication is

Associative: $(\alpha \cdot \beta) \cdot \gamma = \alpha \cdot (\beta \cdot \gamma) \quad \forall \alpha, \beta, \gamma$

Not necessarily commutative	Commutative $\alpha \cdot \beta = \beta \cdot \alpha \ \forall \alpha, \beta \in \mathbb{F}$

$\exists$ identity 1: $\alpha \cdot 1 = 1 \cdot \alpha = \alpha, \forall \alpha.$

$\alpha \in R, \alpha \neq 0 \not\Rightarrow \alpha^{-1}$ exists	$\exists$ Inverse: $\forall \alpha \neq 0$, $\exists$ inverse $a^{-1} \ni$ $\alpha \cdot (\alpha^{-1}) = (\alpha^{-1}) \cdot \alpha = 1$

Distributive law: $\forall \alpha, \beta, \gamma$

D_L: $\alpha \cdot (\beta + \gamma) = \alpha \cdot \beta + \alpha \cdot \gamma$

D_R: $(\beta + \gamma) \cdot \alpha = \beta \cdot \alpha \quad \gamma \cdot \alpha$

From these axioms follow four important facts.

7 Fact. In any ring and in any field, the identities 0 and 1 are unique. (This is easily shown by contradiction.)

8 Fact. In a ring, the cancellation law does not necessarily hold; more precisely, in a ring

$$\left.\begin{array}{l}\alpha\beta = \alpha\gamma \\ \text{and} \\ \alpha \neq 0\end{array}\right\} \quad \text{do not necessarily imply } \beta = \gamma.$$

Example. Consider the noncommutative ring $\mathbb{R}^{2\times 2}$:

$$\begin{bmatrix} 0 & 1 \\ 0 & 0 \end{bmatrix} \cdot \begin{bmatrix} 1 & 1 \\ 0 & 0 \end{bmatrix} = \begin{bmatrix} 0 & 1 \\ 0 & 0 \end{bmatrix} \cdot \begin{bmatrix} 2 & 0 \\ 0 & 0 \end{bmatrix} = \begin{bmatrix} 0 & 0 \\ 0 & 0 \end{bmatrix}$$

$$\alpha \cdot \beta = \alpha \cdot \gamma = 0$$

but clearly $\beta \neq \gamma$.

9 Remark. The cancellation law holds in any field $\mathbb{F}$ because $\alpha \neq 0$

$$\Rightarrow \exists\, \alpha^{-1} \in \mathbb{F} \text{ and } \alpha^{-1}(\alpha\beta) = \alpha^{-1}(\alpha\gamma)$$

$$\Rightarrow (\alpha^{-1}\alpha)\beta = (\alpha^{-1}\alpha)\gamma \qquad \text{(associativity)}$$

$$\Rightarrow \beta = \gamma \qquad (\alpha^{-1}\alpha = 1)$$

10 Remark. We know some rings for which the cancellation law holds: e.g., $\mathbb{Z}$, $\mathbb{R}[s]$, $\mathbb{C}[s]$, $\mathbb{R}_p(s)$, $\mathbb{R}_{p,o}(s)$, $\mathcal{R}(0)$, $\mathcal{R}_o(0)$, $\mathcal{R}_{\mathcal{U}}$. Such rings are called integral domains or, better yet, entire rings.

11 Fact. $\forall \alpha \in \mathcal{R}$, $\alpha \cdot 0 = 0 \cdot \alpha = 0$.

Proof. $\alpha + 0 = \alpha \Rightarrow \alpha \cdot (\alpha + 0) = \alpha \cdot \alpha \Rightarrow \alpha \cdot \alpha + \alpha \cdot 0 = \alpha \cdot \alpha$. Adding $-(\alpha \cdot \alpha)$ to both sides gives $\alpha \cdot 0 = 0$. Repeat the proof but multiply by α on the right: $0 + \alpha = \alpha \Rightarrow (0 + \alpha) \cdot \alpha = \alpha \cdot \alpha$ etc. gives $0 \cdot \alpha = 0$. ∎

12 Fact. $\forall \alpha, \beta \in \mathcal{R}$, $(-\alpha)\beta = -(\alpha \cdot \beta) = \alpha \cdot (-\beta)$.

Proof. $0 = 0 \cdot \beta = [\alpha + (-\alpha)] \cdot \beta = \alpha \cdot \beta + (-\alpha) \cdot \beta \Rightarrow -(\alpha \cdot \beta) = (-\alpha) \cdot \beta$

$0 = \alpha \cdot 0 = \alpha \cdot [\beta + (-\beta)] = \alpha \cdot \beta + \alpha \cdot (-\beta) \Rightarrow -(\alpha \cdot \beta) = \alpha \cdot (-\beta)$. ∎

13 Exercise. Show, from the axioms, that in any ring R,

$$\left(\sum_{i=1}^{n} \alpha_i\right) \cdot \left(\sum_{k=1}^{m} \beta_k\right) = \sum_{i=1}^{n} \sum_{k=1}^{m} \alpha_i \beta_k .$$

Appendix B: Matrices with Elements in a Commutative Ring $\mathbb{K}$

1 Let $\mathbb{K}$ be a commutative ring, i.e., in addition to the standard axioms, we have

2 $$pq = qp \quad \forall p, q \in \mathbb{K}.$$

3 Examples. $\mathbb{K}$ might be

(1) any field: $\mathbb{R}$, $\mathbb{R}(s)$, $\cdots$;
(2) $\mathbb{R}[s]$, $\mathbb{R}_p(s)$, $\mathbb{R}_{p,o}(s)$, $\mathcal{R}(0)$, $\mathcal{R}_0(0)$, $\cdots$;
(3) scalar convolution operators: $p*q = q*p$.

4 Fact. For $n > 1$, $\mathbb{K}^{n\times n}$ is a noncommutative ring.

Indeed: $p \in \mathbb{K}^{n\times n}$ means that $\forall i,j \in \underline{n}$, $p_{ij} \in \mathbb{K}$; and

5 $$(P + Q)_{ij} = p_{ij} + q_{ij}; \; (PQ)_{ij} = \sum_k p_{ik}q_{kj}.$$

Since all these operations are in the ring $\mathbb{K}$, it is easy to verify that all the axioms of rings are satisfied by P and Q.

6 Fact. The definition and properties of determinants hold as in the case of elements in a field (as long as one does not take inverses!).

For example, if $P,Q \in \mathbb{K}^{n\times n}$, then $\det(PQ) = \det(P)\cdot\det(Q)$.

7 Fact [Cramer's rule]. Let $P \in \mathbb{K}^{n\times n}$, hence $\det P \in \mathbb{K}$. Let Adj(P) denote, as usual, the "classical adjoint" of P. By direct calculation we have

8 (a) $\mathrm{Adj}(P)\, P = P\, \mathrm{Adj}(P) = (\det P)I_n$

(b) $P \in \mathbb{K}^{n\times n}$ has an inverse in $\mathbb{K}^{n\times n}$

$\Leftrightarrow$

9 det P has an inverse in $\mathbb{K}$.

In that case,

10 $$P^{-1} = \mathrm{Adj}(P)\,[\det(P)]^{-1} \in \mathbb{K}^{n\times n}.$$

11 Comment. From (8) and (10) it follows that P has a right inverse iff it has a left inverse: the common right and left inverse of P is called the inverse of P.

12 Proof of Fact 7: indications

(a) (8) is equivalent to [Macl.1]

$$\sum_{k=1}^{n} c_{ki}\, p_{kj} = \sum_{k=1}^{n} p_{ik}\, c_{jk} = \delta_{ij}\,[\det(P)] \quad \forall i,j \in \underline{n},$$

where (1) $\delta_{ij} = 0 \ \forall i \neq j$ and $\delta_{ij} = 1 \ \forall i = j$.

(2) c_{ij} is the cofactor of element p_{ij} of P, i.e., $c_{ij} = (-1)^{i+j} m_{ij}$ with m_{ij} the minor of order n - 1 obtained by crossing out row i and column j of P.

(b) If P has an inverse $P^{-1} \in \mathbb{K}^{n\times n}$, then by the axioms of $\mathbb{K}$, $\det(P^{-1}) \in \mathbb{K}$; now $P\,P^{-1} = I_n$ implies $[\det(P)]\,[\det(P^{-1})] = 1$: hence (9) holds. Conversely, if (9) holds, then the RHS of (10) $\in \mathbb{K}^{n\times n}$ and is the inverse of P according to (8). ∎

15 Fact. Let $P \in \mathbb{K}^{m\times n}$, $Q \in \mathbb{K}^{n\times m}$; then

$$PQ \in \mathbb{K}^{m\times m} \quad \text{and} \quad QP \in \mathbb{K}^{n\times n}$$

are well defined by the usual operations (see (5) above).

Here is an extremely useful theorem.

16 Theorem. Let $\mathbb{K}$ be a commutative ring. Let $P \in \mathbb{K}^{n_o\times n_i}$ and $F \in \mathbb{K}^{n_i\times n_o}$. Then,

17 (i) $$(I_{n_o} + PF)^{-1} \in \mathbb{K}^{n_o\times n_o} \Leftrightarrow (I_{n_i} + FP)^{-1} \in \mathbb{K}^{n_i\times n_i}.$$

18 (ii) $$(I_{n_o} + PF)^{-1} = I_{n_o} - P(I_{n_i} + FP)^{-1}F$$

19 $$(I_{n_i} + FP)^{-1} = I_{n_i} - F(I_{n_o} + PF)^{-1}P.$$

20 (iii) Either $(I_{n_o} + PF)^{-1} \in \mathbb{K}^{n_o \times n_o}$ or $(I_{n_i} + FP)^{-1} \in \mathbb{K}^{n_i \times n_i}$

implies

21 $$P(I_{n_i} + FP)^{-1} = (I_{n_o} + PF)^{-1}P \in \mathbb{K}^{n_o \times n_o}.$$

22 (iv) $\det(I_{n_i} + FP) = \det(I_{n_o} + PF)$. ∎

23 Remark: Special case of (ii). If $F = I_n$ and $P \in \mathbb{K}^{n \times n}$, then

24 $$(I_n + P)^{-1} = I_n - (I_n + P)^{-1}P = I_n - P(I_n + P)^{-1}.$$

25 Exercise. Consider the feedback system of Fig. 1, where $G_1 \in \mathbb{R}_p(s)^{n_i \times n_o}$, $G_2 \in \mathbb{R}_p(s)^{n_o \times n_i}$. Calculate the four closed-loop transfer functions $H_{y_j u_i}$: $u_i \mapsto y_j$. Prove that these four $H_{y_j u_i} \in E(\mathbb{R}_p(s)) \Leftrightarrow \det[I + G_1 G_2]$ has an inverse in $\mathbb{R}_p(s)$.

28 Proof of Theorem 16: (a) we handle (i)-(ii) together. Implication ⇒ of (17) and (19) are handled as follows. Assume $(I_{n_o} + PF)^{-1} \in \mathbb{K}^{n_o \times n_o}$. Then

29 $$I_{n_i} - F(I_{n_o} + PF)^{-1}P \in \mathbb{K}^{n_i \times n_i}.$$

We now verify that (29) is the inverse of $(I_{n_i} + FP)$. By calculation in $\mathbb{K}$:

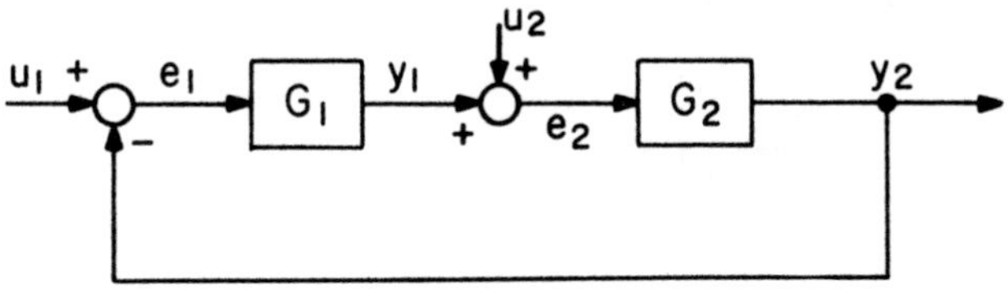

Fig. 1

$$(I_{n_i} + FP)\,[I_{n_i} - F(I_{n_o} + PF)^{-1}P]$$

$$= I_{n_i} + FP - (I_{n_i} + FP)F\,(I_{n_o} + PF)^{-1}P \qquad \text{(distrib. law)}$$

$$= I_{n_i} + FP - F(I_{n_o} + PF)\,(I_{n_o} + PF)^{-1}P \qquad \text{(distrib. law)}$$

$$= I_{n_i}.$$

Hence (19) holds and $(I_{n_i} + FP)^{-1} \in \mathbb{K}^{n_i \times n_i}$ because of (29) and (19).

The implication $\Leftarrow$ of (17) and (18) are handled similarly.

(b) (iii) is established as follows. The distributive law gives

$$(I_{n_o} + PF)P = P + PFP = P(I_{n_i} + FP).$$

Premultiply by $(I_{n_o} + PF)^{-1}$ and postmultiply by $(I_{n_i} + FP)^{-1}$ to obtain equation (21).

(c) We prove (iv). Let

$$M := \left[\begin{array}{c|c} I_{n_i} & F \\ \hline -P & I_{n_o} \end{array}\right] \in \mathbb{K}^{(n_i+n_o)\,(n_i+n_o)}.$$

Now perform on M first $\rho_2 \leftarrow \rho_2 + P\rho_1$, whence M is transformed into

$$M_1 := \left[\begin{array}{c|c} I_{n_i} & F \\ \hline 0 & I_{n_o} + PF \end{array}\right].$$

Perform next on M $\gamma_1 \leftarrow \gamma_1 + \gamma_2\,P$, whence M is transformed into

$$M_2 := \left[\begin{array}{c|c} I_{n_i} + FP & F \\ \hline 0 & I_{n_o} \end{array}\right].$$

Obviously, det M = det M_1 = det M_2, which implies (22). ∎

30 Comment. Appropriate references for Appendices A and B are [Sig.1], [MacL.1], [Jac.1].

Appendix C: Division of a Polynomial Vector on the Left by a Polynomial Matrix

1 Data. Let $D \in \mathbb{R}[s]^{n\times n}$ be a nonsingular polynomial matrix. Let $n \in \mathbb{R}[s]^n$ be a polynomial vector. ■

By Theorem 2.4.3.37.L we know that there exist unique polynomial vectors q and r in $\mathbb{R}[s]^n$ s.t.

2 $$n = Dq + r \quad \text{with} \quad D^{-1} r \in \mathbb{R}_{p,o}(s)^n .$$

We wish to develop an algorithm which computes q and r. We use ideas of [Emr.1], [Fuh.1], and [Kal.1].

Obviously, the division on the right of a row vector $n \in (\mathbb{R}[s]^n)^T$ by a nonsingular polynomial matrix $D \in \mathbb{R}[s]^{n\times n}$ is similar.

3 Lemma. In the division (2), without loss of generality, we may assume in (1) that D is row reduced with positive row degrees (equiv. D^{-1} is strictly proper), and with highest row-degree coefficient matrix $D_h = I$.

Proof. Indeed, when D is not as in the statement of Lemma 3, consider the following algorithm.

1). Find a unimodular matrix $L(s) \in \mathbb{R}[s]^{n\times n}$, s.t.

4 $$[\tilde{D} \mid \tilde{n}] := L[D \mid n],$$

where $\tilde{D}$ is row reduced with row degrees in decreasing order (the latter can be obtained by reordering rows). Store $L(s)^{-1} \in \mathbb{R}[s]^{n\times n}$.

2). Using the highest row-degree coefficient matrix $\tilde{D}_h$ of $\tilde{D}$, obtain

5 $$[\bar{D} \mid \bar{n}] := [\tilde{D}\,\tilde{D}_h^{-1} \mid \tilde{n}],$$

where $\bar{D}$ is now row reduced with $\bar{D}_h = I$; in particular, $\bar{D}$ can be partitioned into

6 $$\bar{D} = \left[\begin{array}{c|c} \bar{D}_{11} & \bar{D}_{12} \\ \hline 0 & I \end{array}\right] \begin{matrix} \}\nu \\ \\ \end{matrix} \in \mathbb{R}[s]^{n\times n},$$
(the first block column has width ν)

where

7 $\bar{D}_{11}$ is row reduced with <u>positive</u> row degrees (equiv. $\bar{D}_{11}^{-1}$ is <u>strictly</u> proper), and highest row-degree coefficient matrix $\bar{D}_{11h} = I$.

3). Using partitions as in (6), let

8 $$\bar{n} = \begin{bmatrix} \bar{n}_1 \\ \hline \bar{n}_2 \end{bmatrix} \in \mathbb{R}[s]^n.$$

4). Find the quotient $\bar{q}_1 \in \mathbb{R}[s]^\nu$ and remainder $\bar{r}_1 \in \mathbb{R}[s]^\nu$ of the division on the left of $\bar{n}_1 - \bar{D}_{12}\,\bar{n}_2 \in \mathbb{R}[s]^\nu$ by $\bar{D}_{11}$, equiv.

9 $$\bar{n}_1 - \bar{D}_{12}\,\bar{n}_2 = \bar{D}_{11}\,\bar{q}_1 + \bar{r}_1 \quad \text{s.t.} \quad \bar{D}_{11}^{-1}\,\bar{r}_1 \in \mathbb{R}_{p,o}(s)^\nu.$$

5). Set

10 $$r := L^{-1}\begin{bmatrix} \bar{r}_1 \\ \hline \theta \end{bmatrix} \in \mathbb{R}[s]^n \quad \text{and} \quad q := \tilde{D}_h^{-1}\begin{bmatrix} \bar{q}_1 \\ \hline \bar{n}_2 \end{bmatrix} \in \mathbb{R}[s]^n.$$

End of Algo.

<u>Claim</u>. The polynomial vectors r and q, given by (10), are the remainder and quotient of the division on the left, (2), of n by D.

Indeed, by (6)-(10),

11 $$\bar{n} = \bar{D}\,\bar{q} + \bar{r} \quad \text{with} \quad \bar{D}^{-1}\,\bar{r} \in \mathbb{R}_{p,o}(s)^n,$$

where

12 $$\bar{q} := \begin{bmatrix} \bar{q}_1 \\ \hline \bar{n}_2 \end{bmatrix} \in \mathbb{R}[s]^n \quad \text{and} \quad \bar{r} := \begin{bmatrix} \bar{r}_1 \\ \hline \theta \end{bmatrix} \in \mathbb{R}[s]^n.$$

Hence by (4)-(5) and (10)-(12),

2 $$n = Dq + r \quad \text{with} \quad D^{-1} r \in \mathbb{R}_{p,o}(s)^n.$$

Therefore, the claim holds and Lemma 3 must hold. Indeed, in the algorithm, division on the left by D, (2), is reduced to division on the left by $\bar{D}_{11}$, (9), where, (7), $\bar{D}_{11}$ has the properties requested in the statement of Lemma 3. ∎

13 Comment. In the algorithm above it may happen that $\bar{D}^{-1}$ is strictly proper. Then in (6), $\bar{D} = \bar{D}_{11}$ and Lemma 3 still holds with obvious modifications to the algorithm.

By Lemma 3, without loss of generality, the division on the left of a polynomial vector $n \in \mathbb{R}[s]^n$ by a nonsingular polynomial matrix can be studied under the following data.

15 Data. We are given a polynomial vector $n \in \mathbb{R}[s]^n$ and a nonsingular matrix $D \in \mathbb{R}[s]^{n \times n}$ which is row reduced with positive row degrees $\rho_i > 0$, $\forall i \in \underline{n}$, and with highest row-degree coefficient matrix $D_h = I$. Hence

16 $$D(s) = \text{diag}[s^{\rho_i}] + D_\ell(s),$$

where

17 $$\rho_i := \partial_{ri}[D] > 0, \quad \forall i \in \underline{n},$$

and

18 $D_\ell(s) \in \mathbb{R}[s]^{n \times n}$ accounts for the terms of degree smaller than ρ_i for every entry in each row of D. ∎

Before giving an algorithm for division on the left, we should observe the following.

20 Lemma. Let $D \in \mathbb{R}[s]^{n \times n}$ be as in (15) and let the set $\mathbb{R}_D$ be defined by

21 $$\mathbb{R}_D := \{x \in \mathbb{R}[s]^n : D^{-1} x \in \mathbb{R}_{p,o}(s)^n\}.$$

U.t.c.

22 (a) $\mathbb{R}_D$ is an $\mathbb{R}$-linear space;

(b) $x \in \mathbb{R}_D$

if and only if

23 $$x(s) = (x_i(s))^T_{i\in\underline{n}},$$

where

$$x_i(s) := x_{i1}\, s^{\rho_i - 1} + x_{i2}\, s^{\rho_i - 2} + \cdots + x_{i\rho_i} \in \mathbb{R}[s], \quad \forall i \in \underline{n};$$

$$\rho_i = \partial_{ri}[D], \quad \forall i \in \underline{n};$$

some of the x_{ij}'s, $j \in \rho_i$, may be zero.

Proof. Exercise. ∎

24 Exercise. Use contradiction to prove that the division (2) leads to a unique quotient and a unique remainder. ∎

Let us now return to the division on the left by $D \in \mathbb{R}[s]^{n\times n}$, where, $\forall x \in \mathbb{R}[s]^n$, there exists a unique quotient q and remainder r s.t.

$$x = Dq + r \quad \text{s.t.} \quad D^{-1}\, r \in \mathbb{R}_{p,o}(s)^n.$$

Hence we can define a quotient and remainder operator $\tilde{q}$, resp. $\tilde{r}$ s.t.

25 $$\tilde{q} : \mathbb{R}[s]^n \to \mathbb{R}[s]^n : x \mapsto \tilde{q}(x),$$

26 $$\tilde{r} : \mathbb{R}[s]^n \to \mathbb{R}_D : x \mapsto \tilde{r}(x),$$

where

27 $$x = D\, \tilde{q}(x) + \tilde{r}(x).$$

28 Exercise. Show that the operators $\tilde{q}$ and $\tilde{r}$ are s.t.,

29 (a) $\tilde{q}$ and $\tilde{r}$ are $\mathbb{R}$-linear operators,

(b) $\forall x(s) \in \mathbb{R}[s]^n$,

30 $$\tilde{r}(sx(s)) = \tilde{r}\big(s\tilde{r}(x(s))\big).$$ ∎

We study now a basic operation for division on the left by D.

35 Lemma. Let D be as in (15). Let $x(s) \in \mathbb{R}_D$ (equiv. x(s) is given by (23)). U.t.c.

36 $$sx(s) = D(s)\,\tilde{q}(sx(s)) + \tilde{r}(sx(s)),$$

where, with (23),

37 $$\tilde{q}(sx(s)) = (x_{i1})^T_{i\in\underline{n}} \in \mathbb{R}^n$$

and

38 $$D(s)^{-1}\,\tilde{r}(sx(s)) \in \mathbb{R}_{p,o}(s)^n \text{ (equiv. } \tilde{r}(sx(s)) \in \mathbb{R}_D).$$

Proof. Clearly from (36)-(37), (23), and (16)-(18),

39 $$\tilde{r}(sx(s)) = sx(s) - D(s)\,\tilde{q}(sx(s)) \in \mathbb{R}[s]^n$$

is s.t., $\forall i \in \underline{n}$, the ith component reads

40 $$x_{i2}\, s^{\rho_i - 1} + x_{i3}\, s^{\rho_i - 2} + \cdots + x_{i\rho_i}\, s - D_{\ell i}(s)\,\tilde{q}(sx(s)),$$

where $D_{\ell i}$ denotes the ith row of $D_\ell \in \mathbb{R}[s]^{n\times n}$ in (16). Clearly, $\forall i \in \underline{n}$, the degree of the polynomial (40) is smaller than ρ_i. Therefore, since D is row reduced as in (15), (38) must hold. ∎

41 Comment. For D as in (15) and for $x \in \mathbb{R}_D$, Lemma 35 shows how to perform the division on the left of sx(s) by D: $\forall i \in \underline{n}$, the ith component of the (constant) quotient vector is the coefficient of s^{ρ_i} of the ith (polynomial) component of sx(s).

We are now able to prove the main theorem of this appendix; its statement contains the division algorithm.

45 Theorem [Division on the left of a polynomial vector n by a polynomial matrix D]. Let $D \in \mathbb{R}[s]^{n\times n}$ be as in (15) and let $n \in \mathbb{R}[s]^n$ be given by

46 $$n(s) = n_o\, s^k + n_1\, s^{k-1} + \cdots + n_k,$$

where $n_t \in \mathbb{R}^n$, $\forall t = 0, 1, \cdots, k$.

Consider the quotient and remainder operators $\tilde{q}$, resp. $\tilde{r}$, as defined in (25)-(27). Now consider the discrete-time system Σ described by the equations

47[a] $$x_{t+1}(s) = \tilde{r}(sx_t(s)) + n_t,$$

$$\forall t = 0, 1, 2, \cdots, k,$$

47[b] $$y_{t+1} = \tilde{q}(sx_t(s)),$$

where, $\forall t = 0, 1, \cdots, k$, $n_t \in \mathbb{R}^n$ is the coefficient vector of s^{k-t} in (46), and

48 $$x_0(s) := \theta \in \mathbb{R}_D.$$

U.t.c.

49 (a) $\forall t = 1, 2, \cdots, k+1$, $x_t(s) \in \mathbb{R}_D$ and $y_t \in \mathbb{R}^n$.

(b) The remainder and quotient of the division on the left of n by D are given by

50 $$\tilde{r}(n(s)) = x_{k+1}(s) \in \mathbb{R}_D,$$

51 $$\tilde{q}(n(s)) = \sum_{t=1}^{k} y_{t+1}\, s^{k-t} \in \mathbb{R}[s]^n.$$

52 Comments. (a) The remainder is the state of Σ at $t = k + 1$ and the quotient is given by the outputs y_t for $t = 2, \cdots, k+1$.

(b) At each iteration we divide $sx(s)$ by D with $x \in \mathbb{R}_D$.

53 Proof of Theorem 45

(a): This follows from (47)-(48) and Lemma 35. Note with care that $\mathbb{R}_D$ is an $\mathbb{R}$-linear space and $n_t \in \mathbb{R}_D$.

(b) (50) is established by induction, using (29)-(30).

Indeed:

$$x_1(s) = n_o = \tilde{r}(n_o)$$
$$x_2(s) = \tilde{r}(s\tilde{r}(n_o)) + n_1 = \tilde{r}(sn_o) + \tilde{r}(n_1) = \tilde{r}(sn_o + n_1)$$
$$x_3(s) = \tilde{r}(s\tilde{r}(sn_o + n_1)) + n_2 = \tilde{r}(s^2 n_o + sn_1 + n_2)$$
$$\cdots\cdots$$
$$x_{k+1}(s) = \tilde{r}(s^k n_o + s^{k-1} n_1 + \cdots + n_k) = \tilde{r}(n(s)).$$

(51) follows also by induction.

Indeed:

$$sn_o + n_1 = D(s)\,\tilde{q}(sn_o) + \tilde{r}(sn_o + n_1)$$

$$= D(s)\,y_2 + x_2(s)$$

$$s^2 n_o + sn_1 + n_2 = s(sn_o + n_1) + n_2$$

$$= s(D(s)\,y_2 + x_2(s)) + n_2$$

$$= sD(s)\,y_2 + sx_2(s) + n_2$$

$$= D(s)(sy_2 + y_3) + x_3(s)$$

...

$$n(s) = s^k\, n_o + s^{k-1}\, n_1 + \cdots + n_k$$

$$= D(s)\,\Big(\sum_{t=1}^{k} y_{t+1}\, s^{k-t}\Big) + x_{k+1}(s)$$

$$= D(s)\,\tilde{q}(n(s)) + \tilde{r}(n(s)).$$

∎

55 <u>Example</u>. Let

$$D(s) = \left[\begin{array}{c|c} (s+1)^2(s+2) & s+2 \\ \hline 0 & (s+2)^2 \end{array}\right],\quad n(s) = \left[\begin{array}{c} s^4 \\ \hline s^2 \end{array}\right].$$

Then

$$\tilde{q}(n(s)) = \left[\begin{array}{c} s-4 \\ \hline 1 \end{array}\right] \text{ and } \tilde{r}(n(s)) = \left[\begin{array}{c} 11s^2 + 17s + 6 \\ \hline -4s-4 \end{array}\right].$$

∎

60 <u>Final Comment</u>: <u>Division of a Polynomial Matrix $N \in \mathbb{R}[s]^{n\times p}$ on the Left by a Nonsingular Polynomial Matrix</u> $D \in \mathbb{R}[s]^{n\times n}$. This division is carried out by successively dividing on the left the columns of N by D.

References

Cal. 1 F. M. Callier and C. A. Desoer, "Stabilization, Tracking and Disturbance Rejection in Multivariable Convolution Systems," Annales de la Société Scientifique de Bruxelles, T. 94, I, pp. 7-51, 1980.

Cal. 2 F. M. Callier, V. H. L. Cheng, and C. A. Desoer, "Dynamic Interpretation of Poles and Transmission Zeros for Distributed Multivariable Systems," IEEE Trans. Circ. and Syst., Vol. CAS-28, pp. 300-307, 1981.

Che. 1 C. T. Chen, "Introduction to Linear System Theory," Holt, Rinehart and Winston, New York, 1970.

Chen. 1 M. J. Chen, C. A. Desoer, and G. F. Franklin, "Algorithmic Design of Single-Input Single-Output Systems with a Two-Input One-Output Controller," Memorandum Electronics Research Laboratory, University of California, Berkeley, No. UCB/ERL M81/24, July 1921.

Cru. 1 J. B. Cruz, J. S. Freudenberg, and D. P. Looze, "A Relation between Sensitivity and Stability of Multivariable Feedback Systems," IEEE Trans. Auto. Control, Vol. AC-26, pp. 66-74, February 1981.

Dav. 1 E. J. Davison, "The Robust Control of a Servomechanism Problem for Linear Time-Invariant Multivariable Systems," IEEE Trans. Auto. Control, Vol. AC-21, pp. 25-34, February 1976.

Des. 1 C. A. Desoer and J. D. Schulman, "Zeros and Poles of Matrix Transfer Functions and their Dynamical Interpretations," IEEE Trans. Circ. and Syst., Vol. CAS-21, January 1974.

Des. 2 C. A. Desoer and Y. T. Wang, "Linear Time-Invariant Robust Servomechanism Problem: A Self-Contained Exposition," in "Advances in Control and Dynamical Systems," Vol. 16, C. T. Leondes (Ed.), Academic Press, New York, pp. 81-129, 1980.

Des. 3 C. A. Desoer and Y. T. Wang, "Foundations of Feedback Theory for Nonlinear Dynamical Systems," IEEE Trans. Circ. and Syst., Vol. CAS-27, pp. 320-323, April 1980.

Des. 4 C. A. Desoer and M. Vidyasagar, "Feedback Systems: Input-Output Properties," Academic Press, New York, 1975.

Des. 5 C. A. Desoer and M. J. Chen, "Design of Multivariable Feedback Systems with Stable Plants," IEEE Trans. Auto. Control, Vol. AC-26, 2, pp. 408-415, April 1981.

Doy. 1 J. C. Doyle and G. Stein, "Multivariable Feedback Design: Concepts for a Classical/Modern Synthesis," IEEE Trans. Auto. Control, Vol. AC-26, pp. 4-17, February 1981.

Emr. 1 E. Emre, "The Polynomial Equation $QQ_c + RP_c = \Phi$ with Application to Dynamic Feedback," SIAM Jour. Contr. and Opt., Vol. 18, pp. 611-620, November 1980.

Fra. 1 B. A. Francis and M. Vidyasagar, "Algebraic and Topological Aspects of the Servo Problem for Lumped Linear Systems," S & IS Report No. 8003, Yale University, New Haven, Conn., 1980.

Fuh. 1 P. A. Fuhrmann, "Algebraic System Theory: An Analyst's Point of View," Jour. Franklin Inst., Vol. 301, pp. 521-540, 1976.

Gan. 1 F. R. Gantmacher, "The Theory of Matrices," Vol. 1, Chelsea, New York, 1959.

Gar. 1 B. S. Garbow, M. M. Boyle, J. J. Dongarra, and C. B. Moler, "Matrix Eigensystem Routines -- EISPACK Guide Extension," Lecture Notes in Computer Science, No. 51, Springer Verlag, New York, 1977.

Goh. 1 I. Gohberg and L. Rodman, "On Spectral Analysis of Non-Monic Matrix and Operator Polynomials, I. Reduction to Monic Polynomials," Israel Jour. Math., Vol. 30, pp. 133-151, 1978.

Goh. 2 I. Gohberg and L. Rodman, "On Spectral Analysis of Non-monic Matrix and Operator Polynomials, II. Dependence on the Finite Spectral Data," Israel Jour. Math., Vol. 30, No. 4, pp. 321-334, 1978.

Hun. 1 N. T. Hung and B. D. O. Anderson, "Triangularization for the Design of Multivariable Control Systems," IEEE Trans. Auto. Contr., Vol. AC-24, pp. 455-460, 1979.

Jac. 1 N. Jacobson, "Basic Algebra I," Freeman, San Francisco, 1974.

Kai. 1 T. Kailath, "Linear Systems," Prentice-Hall, Englewood Cliffs, N. J., 1980.

Kal. 1 R. E. Kalman, P. L. Falb, and M. A. Arbib, "Topics in Mathematical System Theory," McGraw-Hill, New York, 1969.

Kuc. 1 V. Kucera, "Discrete Linear Control, The Polynomial Approach," John Wiley, New York, 1979.

MacD. 1 C. C. MacDuffee, "The Theory of Matrices," Springer-Verlag, Berlin, 1933, reprinted Chelsea, New York, 1956.

MacL. 1 S. MacLane and G. Birkhoff, "Algebra," 2nd. ed., Mcmillan, New York, 1979 (1st. ed., 1967).

Nob. 1 B. Noble and J. W. Daniel, "Applied Linear Algebra," Prentice-Hall, Englewood Cliffs, N. J., 1977, rev. ed. (1st. ed., 1969).

Per. 1 L. Pernebo, "An Algebraic Theory for the Design of Controllers of Linear Multivariable Systems, Part I: Structure Matrices and Feedforward Design, Part II: Feedback Realizations and Feedback Design," IEEE Trans. Auto. Control, Vol. AC-26, pp. 171-193, February 1981.

Pol. 1 E. Polak, "Optimization-based Computer-aided Design of Control Systems," Proceedings of the 1981 Joint Automatic Control Conference.

Pug. 1 A. C. Pugh and P. R. Ratcliffe, "On the Zeros and Poles of a Rational Matrix," Int. Jour. Control, Vol. 30, pp. 213-226, 1979.

Ros. 1 H. H. Rosenbrock, "State Space and Multivariable Theory," Wiley, New York, 1970.

Ros. 2 H. H. Rosenbrock and G. E. Hayton, "The General Problem of Pole Assignment," Int. Jour. Control, Vol. 27, No. 6, pp. 837-852, 1978.

Rud. 1 W. Rudin, "Real and Complex Analysis," McGraw-Hill, New York, 1974.

Saf. 1 M. G. Safonov, A. J. Laub, and G. L. Hartmann, "Feedback Properties of Multivariable Systems: The Role and Use of the Return Difference Matrix," IEEE Trans. Auto. Control, Vol. AC-26, pp. 47-65, February 1981.

Saf. 2 M. G. Safonov, "Stability and Robustness of Multivariable Feedback Systems," MIT Press, Cambridge, Mass., 1980.

Sig. 1 L. E. Sigler, "Algebra," Undergraduate Texts in Mathematics, Springer Verlag, New York, 1976.

Ste. 1 G. W. Stewart, "Introduction to Matrix Computations," Academic Press, New York, 1973.

Tem. 1 G. C. Temes and J. W. La Patra, "Circuit Synthesis and Design," McGraw-Hill, New York, 1977.

Ver. 1 G. Verghese, "Infinite-Frequency Behavior in Generalized Dynamical Systems," Ph.D. Thesis, Stanford University, Stanford, Calif., December 1978.

Vid. 1 M. Vidyasagar, H. Schneider, and B. A. Francis, "Algebraic and Topological Aspects of Feedback Stabilization," Tech., Report No. 80-09, Dept. of Electrical Engineering, University of Waterloo, September 1980.

Vid. 2 M. Vidyasagar, "Nonlinear System Analysis," Prentice-Hall, Englewood Cliffs, N. J., 1978.

Vid. 3 M. Vidyasagar, "Input-Output Analysis of Large-Scale Interconnected Systems," Lecture Notes in Control and Information Sciences, Springer Verlag, New York, 1981.

Wil. 1 J. C. Willems, "The Analysis of Feedback Systems," MIT Press, Cambridge, Mass, 1971.

You. 1 D. C. Youla, J. J. Bongiorno, Jr., and C. N. Lu, "Single-Loop Feedback Stabilization of Linear Multivariable Dynamical Plants," Automatica, Vol. 10, pp. 159-173, 1974.

Zam. 1 G. Zames, "Feedback and Optimal Sensitivity: Model Reference Transformations, Multiplicative Seminorms and Approximate Inverses," IEEE Trans. Auto. Control, Vol. AC-26, pp. 301-320, April 1981.

Symbols

I. Sets

$\in$	$a \in A$	a is an element of A; a belongs to A
$\subset$	$A \subset B$	set A is contained in set B; A is a subset of B
$\cup$	$A \cup B$	union of set A with set B
$\cap$	$A \cap B$	intersection of set A and set B
$\Rightarrow$	$p \Rightarrow q$	p implies q; equivalently, "not q" implies "not p"
$\Leftarrow$	$p \Leftarrow q$	q implies p
$\Leftrightarrow$	$p \Leftrightarrow q$	p if and only if q; equivalently, p implies q and q implies p
	$\mathring{M}$	interior of the set M
	$\bar{M}$	closure of the set M (Note: For $z \in \mathbb{C}$, $\bar{z}$ denotes the complex conjugate of z)
	M^C	complement of the set M
:=	A := B	the set A is by definition the set B
=:	A =: B	the set B is by definition the set A
\	A \ B	the set difference of set A minus set B, equiv. $A \cap B^C$
$\times$	$A \times B$	the cartesian product of set A times set B; equiv. $A \times B = \{(a, b); a \in A, b \in B\}$
	ϕ	the empty set

II. Algebra

$\mathbb{C}$	field of complex numbers
$\underline{k}$	the set of integers $\{1, 2, \cdots, k\}$
$\mathbb{N}$	set of nonnegative integers, namely, $\{0, 1, 2, \cdots\}$
$\mathbb{R}$	field of real numbers
$\mathbb{Z}$	ring of integers, namely, $\{\cdots, -1, 0, 1, 2, \cdots\}$
$\mathbb{Q}$	field of rational numbers
R	a ring
$\mathbb{K}$	a commutative ring
$\mathbb{F}$	a field
$\mathbb{F}[x]$	ring of polynomials in one variable with coefficients in the field $\mathbb{F}$ (e.g., $\mathbb{R}[s]$, $\mathbb{C}[s]$)

$\mathbb{F}(x)$	field of rational functions in one variable with coefficients in $\mathbb{F}$ (e.g., $\mathbb{R}(s)$, $\mathbb{C}(s)$)
$\mathbb{R}_p(s)$, $\mathbb{R}_{p,o}(s)$	the ring of proper (equiv. bounded at infinity), resp. strictly proper (equiv. zero at infinity), rational functions in s with coefficients in $\mathbb{R}$
$\mathbb{C}_+$	$:= \{s \in \mathbb{C} : \text{Re } s \geq 0\}$, equiv. the closed right-half of the complex plane
$\mathcal{U}$	an undesirable subset of $\mathbb{C}$, which is symmetric w.r.t. the real axis and contains $\mathbb{C}_+$
$\mathcal{R}(0)$, $\mathcal{R}_o(0)$	the subring of elements of $\mathbb{R}_p(s)$, resp. $\mathbb{R}_{p,o}(s)$, that are analytic in $\mathbb{C}_+$ (equiv. with no poles in $\mathbb{C}_+$)
$\mathcal{R}_{\mathcal{U}}$	the subring of elements of $\mathcal{R}(0)$ that are analytic in $\mathcal{U}$
$(x_k)_{k\in K}$	family of elements; K := index set; the family is a map $K \to X$
A^n	set of n-tuples of elements belonging to the set A (e.g., $\mathbb{R}^n$, $\mathbb{R}[s]^n$, $\mathbb{R}(s)^n$, $\cdots$)
$A^{p\times q}$	set of $p \times q$ arrays of elements belonging to the set A; equiv. the set of $p \times q$ matrices with entries in the set A, (e.g., $\mathbb{R}^{p\times q}$, $\mathbb{R}[s]^{p\times q}$, $\mathbb{R}(s)^{p\times q}$, $\cdots$)
$A \in \mathcal{E}[B]$	matrix A has entries in the set B
∂a	the degree of the polynomial a
$\partial_{ri}[A]$	the ith row degree of the polynomial matrix A
$\partial_{cj}[A]$	the jth column degree of the polynomial matrix A
$\partial_M[A]$	the Mcmillan degree of the rational matrix A
$p\|q$	the polynomial p divides the polynomial q without remainder, or equiv. p is a factor of q
$A \sim B$	the polynomial matrix A is equivalent to the polynomial matrix B, or equiv. there exist unimodular polynomial matrices L and R such that $A = LBR$
$a \sim b$	the polynomial a is equivalent to the polynomial b, or equiv. there exists a nonzero constant k s.t. $a = kb$
a^*, A^*	the complex conjugate transpose of the complex vector a, resp. matrix A
a^T, A^T	the transpose of the vector a, resp. matrix A
ρ_i	the ith row of a matrix
γ_j	the jth column of a matrix
$\mathcal{V}$	the vector space $\mathcal{V}$ (also called linear space)
$\mathcal{M}$	the module $\mathcal{M}$

θ	the zero element of a vector space or module
$N(A)$	the null space of matrix A (or the linear map A)
$R(A)$	the range or image of the matrix A (or the linear map A)
$rk(A)$	the rank of matrix A
$\sigma(A)$	the spectrum of matrix A (equiv. the set of eigenvalues of A)
$\sigma_{max}[A]$, $\bar{\sigma}[A]$	the largest singular value of matrix A
$\sigma_{min}[A]$	the smallest singular value of matrix A
$V_1 \oplus V_2$, $M_1 \oplus M_2$	the direct sum of two vector spaces, resp. modules
$V_1 \perp V_2$	vector space V_1 is orthogonal to vector space V_2
$V_1 \overset{\perp}{\oplus} V_2$	the direct orthogonal sum of vector spaces V_1 and V_2

III. Analysis

$f : A \to B$	f is a function or map, mapping a domain A into the codomain B; also noted $x \mapsto f(x)$, i.e., f is a function which associates with $x \in A$ the image $f(x) \in B$
dom[f]	the domain of the map f
codom[f]	the codomain of the map f
$R[f]$	the range or image of the map f
$N[f]$	the nullspace of the map f
$\exists$	there exists
$\exists!$	there exists a unique element
s.t.	such that
$\mathbb{C}_+$	$:= \{s \in \mathbb{C} : \text{Re } s \geq 0\}$
$\mathring{\mathbb{C}}_-$	$:= \{s \in \mathbb{C} : \text{Re } s < 0\}$, equiv. the open left-half of the complex plane
$\mathring{\mathbb{C}}_+$	$:= \{s \in \mathbb{C} : \text{Re } s > 0\}$
p	the differentiation operator, e.g., $pf = df/dt$
C	the space of continuous functions
C^k	the space of k-times continuously differentiable functions
C^∞	the space of infinitely continuously differentiable functions
$\Vert f \Vert$	the norm of f
$P[f]$, $P[H]$	the set of poles of the vector function f, resp. the matrix function H
$Z[f]$, $Z[H]$	the set of zeros of the vector function f, resp. the matrix function H

$(x_\lambda)_{\lambda\in L}$	the family of elements x_λ with index set L, (special case: sequences: $(x_\lambda)_{\lambda\in \mathbb{N}}$ is often written $(x_\lambda)_0^\infty$ or simply (x_λ)).
$\hat{f}$, $\hat{H}$	the Laplace transform of the vector function f, resp. matrix function H (also denoted by $\mathcal{L}[f]$, resp. $\mathcal{L}[H]$)
$\mathcal{L}^{-1}[\hat{f}]$, $\mathcal{L}^{-1}[\hat{H}]$	the inverse Laplace transform of the vector function $\hat{f}$, resp. matrix function $\hat{H}$.

List of Abbreviations

PMD	polynomial matrix system description
equiv.	equivalently
s.v.d.	singular value decomposition
s.t.	such that
exp. stable	exponentially stable
I/O map	input-output map
r.d.	right divisor
ℓ.d.	left divisor
ℓ.m.	left multiple
r.m.	right multiple
g.c.r.d.	greatest common right divisor
g.c.ℓ.d.	greatest common left divisor
c.r.d.	common right divisor
c.ℓ.d.	common left divisor
r.c.	right coprime
ℓ.c.	left coprime
e.o.	elementary operation
e.r.o.	elementary row operation
e.c.o.	elementary column operation
e.m.	elementary matrix
ℓ.e.m.	left elementary matrix
r.e.m.	right elementary matrix
r.f.	right fraction
r.c.f.	right coprime fraction
ℓ.f.	left fraction
ℓ.c.f.	left coprime fraction
SMM-form	Smith-Mcmillan form
c.r.	column reduced
r.r.	row reduced
p. suff. diff.	piecewise sufficiently differentiable
z-i p-s trajectory	zero-input pseudo-state trajectory
z-i response	zero-input response
z-s p-s trajectory	zero-input pseudo-state trajectory

z-s response	zero-state response
i-d zero	input-decoupling zero
o-d zero	output-decoupling zero
r.c.r.	row-column reduced
r.ℓ.f.	right-left fraction
r.ℓ.c.f.	right-left coprime fraction
int. pr.	internally proper
ℓ.r.f.	left-right fraction
ℓ.r.c.f.	left-right coprime fraction
char. poly.	characteristic polynomial
eqn.	equation
RHS	right-hand side
LHS	left-hand side
comp. eqn.	compensator equation

Index

algorithm
  for coprime factorization, 57
  for decoupling compensator, 230
  for division of vector matrix, 253
  for eigenvalue placement, 192
  for g.c.d. extraction, 51
  for I/O map, 176
  for Hermite form, 34
  for polynomial matrices, 71, 253
  for robust as.tracking s.i.s.o., 179
  for robust as. tracking, 213
  for Smith form, 38
  for Smith Mcmillan form, 63
assumption
  IS, 141
  PMD, 144
  PMD', 153
  WF, 150
Bezout identity, 54, 60, 61, 184, 214
cancellation
  pole-zero, 157
  see also hidden modes
char. poly.
  characteristic polynomial
characteristic polynomial
  of feedback system, 155, 159, 167
  of interconnected system, 148, 149
  of PMD, 94
c.ℓ.d.
  common left divisor
column powers, 116
compensation
  two-step — theorem, 232, 237
comp. eqn.
  compensator equation
compensator equation
  design pt. of view, 192
  existence of solutions, 190
  tracking, 213
    polynomial solutions, 215
    internally proper sol., 216
      existence of, 217
computer-aided design, 244
c.r.
  column reduced
Cramer's rule, 249
c.r.d.
  common right divisor
desensitization
  bounds on, 227, 228
determinant, 249
  property, 250, 251
differential equation
  exponentially stable, 122
  pseudo-state trajectory, 49
  solution space, 30, 42
  state, 43, 49
  well-formed, 109
discrete-time systems, 243
disturbance rejection, 207
distributed systems, 243
distributive laws, 246
division
  theorem, 257

dynamical interpretation
  of exp. stability
    of PMD's, 124-127
    of transfer functions, 127
  of poles, 82
  of pole at ∞, 113
  of zeros, 82
e.c.o.
  elementary column operations
eigenvalue
  closed-loop versus open-loop, 159
  of PMD, 94, 106
    distinct, 130
  placement problem, 181 et seq.
elementary
  matrices, 27, 28
  operations
    on matrices, 26
    on differ. eqn., 29
e.m.
  elementary matrix
e.o.
  elementary operation
eqn.
  equation
equiv.
  equivalently
e.r.o.
  elementary row operation
exp. stable
  exponentially stable
field, 246
g.c.ℓ.d.
  greatest common left divisor
g.c.r.d.
  greatest common right divisor
hidden modes, 104
  unstable, 128, 144, 203, 223, 225
i-d zero
  input decoupling zero
input
  exogeneous, 142
  reference, 172, 199
I/O map
  input-output map
I/O maps
  achievable, 171
    algorithm for, 176
Jordan chain relation, 129
large loop gain, 12
ℓ.c.
  left coprime
ℓ.c.f.
  left coprime fraction
ℓ.d.
  left divisor
left fraction
  int. proper, 137
left-right fraction, 132
  — coprime —, 132
ℓ.e.m.
  left elementary matrix
ℓ.f.
  left fraction
ℓ.m.
  left multiple
ℓ.r.c.f.
  left right coprime fraction
ℓ.r.f.
  left right fraction
int. proper
  internally proper
matrix
  biproper, 116, 118, 182
  gain, 143
  highest degree coefficient, 116

matrix (cont'd)
  indicator ellipsoid, 8
  norm of, 9
  return difference, 11, 159
  rational, rank, 25
  singular value, 7
  singular value decomposition, 2
  see also polynomial matrix, rational transfer for matrix
module, 21
  basis of, 21
  free, 21
  product, 21
  submodule, 21
Nyquist
  test, 160
o-d zero
  output decoupling zero
parametrization
  global, 185
  of internally proper solution, 193
  see also Q-parametrization
PMD
  polynomial system descr., 18
  closed-loop input-output, 146
  complete response, 96
  dynamics of, 92
  exponentially stable, 121, 123
    properties of, 124-127
  minimal, 104
  observability of, 101 et seq.
    complete --, 103
  open-loop, 146, 153
  order of, 92
  pole of, 105
  pseudo state, 92
  transfer function, 95
PMD (cont'd)
  state at t = 0, 93
    normalized, 93
  well-formed, 107 et seq.
    test for, 120
  zero of, 105
  zero-input
    p-s trajectory, 97, 102
      reachable, 97
      unobservable, 102
    well-formed, 111
  zero-state
    p-s trajectory, 94
    response, 95
    well-formed, 113, 114, 116
polynomial
  characteristic
    of feedback system, 155, 159, 167
    of interconnected system, 148, 149
  Euclidian division, 19
  see also, polynomial matrix, PMD, characteristic polynomial
polynomial matrix
  column-degree, 67
  column-reduced, 68, 116
  coprime, left —, right —, 24
  determinantal divisors, 38
  divisors, left —, right —, 24
  elementary operations on —, 26, 27
  equivalence, 29, 41
  greatest common divisor, 24
    extraction of, 51
  Hermite form, 33, 36
  highest degree coeff. matrix, 69, 70
  invariant polynomials, 37
  multiples, right —, left —, 24
  nonsingular, 23
  normal column rank, 25

polynomial matrix (cont'd)
normal row rank, 25
rank, normal-, local-, 116
rank test, 54
row-column reduced, (r.c.r.), 116
row-degree, 69
row-reduced, 68, 116
Smith form, 37
unimodular, 23
problem
compensator, 181
solution, 184 et seq.
int. proper solutions, 187, 190
design of, 192-193
tracking, 196 et seq.
see also tracking
properness
theorem, 170, 223
pseudo-state, 167
p.suff. diff.
piecewise sufficiently differentiable
Q-parametrization, 219 et seq.
formulas, 221
Q defined, 220
theorems, 222 et seq.
limitations on I/O map, 225, 226
global parametrization, 223
well-posedness, 222
design trade-off, 224
decoupling, 229 et seq.
algorithm for, 230
quotient, 256
rank
column —, 25
local, 25
normal, 25
row —, 25
rank test, 54
rational transfer fn. matrix
Bezout identity, 60
biproper, 133, 134
denominator of, 66
left fraction, 56
numerator of, 66
proper, 67
strictly proper, 67
pole, 73
poles and zeros, 75, 79, 80, 82
poles and zeros
dynamical interpretation, 82
realization, 87
right-fraction, 56
Smith-Mcmillan form, 62
zero, 73, 85
reachability
complete- of PMD, 100
of PMD, 96 et seq.
of z-i p-s trajectory, 97
test for, 97
r.c.
right coprime
r.c.f.
right coprime fraction
r.c.r.
row-column reduced
r.d.
right divisor
regulator, 200
r.e.m.
right elementary matrix
remainder, 256
RHS
right hand side
r.f.
right fraction

right fraction
  int. proper, 138
right-left fraction, 131
    exp. stable, 131
    internally proper, 132, 135
ring, 246
  commutative, 246, 249
  Cramer's rule, 249
  euclidian, 18
  of exp. stable tsfer fns, 19
  properties, 247
r.ℓ.c.f.
  right left coprime fraction
r.ℓ.f.
  right left fraction
r.m.
  right multiple
row powers, 116
r.r.
  row reduced
sensitivity, 11
  see also desensitization
  servomechanism, 201
  see also tracking
SMM-form
  Smith-Mcmillan form
s.t.
  such that
stability
  exponential, 122, 151, 168
  robust, 161
  $\mathcal{U}$-stability, 164, 169, 245
  see also characteristic polynomial
summing node, 141
s.v.d.
  singular value decomposition
system
  interconnected, 140
    exponentially stable, 149
  well posed, 222
  feedback
    exp. stability, 153, 154
    single-input-single-output, 163
      et seq.
system matrix, 105
tracking
  asymptotic
    definition, 199
    theory of, 197 et seq.
    necess. cond. for, 201
    suff. cond. for, 204, 206
    robustness of, 207
  compensator problem, 212, 213, 215,
    216, 217
    algorithm, 213
tracking s.i.s.o.
  robust asymptotic, 172 et seq.
    algorithm for, 179
trajectory
  z-i p-s — reachable, 97
  z-i p-s — unobservable, 102
transfer function
  closed-loop, 144
  exp. stable, 19, 127, 128
    properties of, 127
  forward-path, 201
  input-error, 143
  input-output, 143
  open-loop, 143
zero, 73, 85
  at ∞, 122
  blocking zero, 208
  $\mathbb{C}_+$-zero of plant, 160, 225
  decoupling, 128, 152

zero (cont'd)
  input-decoupling, 101
  of PMD, 105
  output-decoupling, 104
z-i p-s trajectory
  zero-input pseudo-state trajectory
z-i response
  zero input response
z-s p-s trajectory
  zero-state pseudo-state trajectory
z-s response
  zero-state response